Klaus Schuster

11 Management Sins That You Should Avoid

Klaus Schuster

11 Management Sins That You Should Avoid

How top managers are killing themselves to have it all: career, relationships and fun

REDLINE | VERLAG

Deutsche Nationalbibliothek Bibliographical Information
The Deutsche Nationalbibliothek lists this publication in the Deutsche Nationalbibliografie. Detailed bibliographical information is available online at **http://d-nb.de**.

Für Fragen und Anregungen:
schuster@redline-verlag.de

1st Edition 2014

Editorial: Pia Gelpke, wortvollendet, Wiesbaden
Translation: Shelley Steinhorst, Düsseldorf
Cover Illustrations: Jarzina Kommunikations-Design, Holzkirchen
Typesetting: Georg Stadler, München
Printing: Books on Demand GmbH, Norderstedt
Printed in Germany

ISBN Print 978-3-86881-562-7
ISBN E-Book (EPUB, Mobi): 978-3-86414-679-4

Table of Contents

Preface

"There is no trap so deadly, as the trap you set for yourself."

Raymond Chandler

What is it that makes some managers so good? Is it their exceptional talents and skills, a good nose for business, a secret recipe or a gut feeling that never fails? Or maybe it's just the right touch, or the right connections?

Like many others, as I began my climb up the ladder, I also thought (with no small amount of vanity) that top managers had to be something very special. And *ergo* that they must be able to do something very special. Since then I've had many opportunities to observe them up close. As a board member of international banks, setting up and winding down financial institutions in Eastern Europe, and now, today, in my own company which advises, coaches and trains managers in all industries and at all levels, I've come into contact with many top executives in many countries—contact with my American colleagues, though infrequent, is often rather comical. It usually starts with the introduction:

"Hey guy, where are you from?"

"Austria"

"Ah! I know. Kangaroos and crocodiles, right?"

"No, that's Australia. I'm from Austria!"

"Yeah man, I know. Sydney and Bondi beach. Fantastic!"

"Noooo! Niki Lauda, Mozart and Arnold Schwarzenegger!"

"Huh? ..."

In spite of these little cross-cultural misunderstandings, I still held on to my belief that top managers really were "top"—I wonder, now, how I could have been so naïve, because what makes top managers so outstanding is not their outstanding qualities.

Even colleagues who have had less success usually have at least one remarkable talent, otherwise they wouldn't have kept their positions for so long. Average managers can have excellent talents. But why aren't these talents helping them more? A good example came recently from the stock market: A lot of traders were making millions—for years—and then one fine day in autumn, a mortgage crisis erupted and the less clever traders suffered one loss after the other. They lost their jobs, dragged their bosses down, and sometimes even ruined their companies in the process. Which just goes to show:

The secret of top-notch managers: they make fewer, and less serious, mistakes.

This reminds me of a board member who, although he was smart enough and had a proven track record to go with it, still wasn't considered a top manager by his peers. He made one fatal mistake: He didn't do enough to improve business. He waited patiently each day for the phenomenal, ground-breaking success to come knocking on his door, thus ensuring his place of glory in the annals of the company for all time. The other board members called him the "tick-tock manager".

We all make mistakes now and then—don't worry. Weak managers, however, don't just make a few mistakes, they commit sins: systematic, chronic, colossal management sins.

That's why they don't manage to achieve phenomenal success, or the long wished for breakthrough. They miss out on that decisive career step or the admiration and respect of their line managers and colleagues. They also lack the esteem of their clients, unshakeable self-confidence and the adoration of their partners and children. They fail because they commit critical management sins. Looking back, I'm almost sure that the aforementioned board member knew, or at least suspected, what fatal mistake he was making, how mad colleagues were with him and that he was, in effect, digging his own management grave.

But what could he have done about it? General management training is often exactly that: general. It doesn't always offer a quick and effective remedy for this type of tick-tock management or the other management sins. In practice, most training measures for managers don't help when they are based on the misconception of ideal management virtues (that have nothing to do with reality and won't help managers keep their white collars clean).

For a long time I brooded over this (seemingly) built in tendency towards self-destruction. Then I remembered my almost twenty years of experience in management jobs and executive board rooms, my own youthful cardinal and deadly sins and I realized: when you consider that general management trainers, bosses, management gurus, editorial journalists, and coaches have been preaching to top managers for decades about the "right" way of doing things, and then aren't even able to do it "right" themselves, then it's probably not in spite of, but because of, this overly-positive didactical style.

This well-meaning advice comes across as too saccharine for most managers tastes. It hasn't been able to prevent me from committing the classic management sins at the executive board room level (and it hasn't made the managers who I've been advising and coaching for years become more virtuous). When this realization hit me, I decided to turn the tables.

From then on I began telling managers what they should *not* be doing. I told them—without naming names, of course—about manager colleagues who were suffering the most because of their *sins*. And, oh, what joy: It worked! And to those who are reading just to gloat over another manager's mistakes, to them I say: go on and gloat—the happier you are, the better you learn. Everyone knows: we learn from our mistakes. But the managers learning the most are the ones learning from the mistakes of others.

In consulting sessions since then, I have definitely seen the metaphorical light going on. International trade journals promptly asked me to publish my findings and I began to enjoy great popularity among German-speaking managers across Europe—at least those who regularly read my columns or invited me to speak. The others kept asking me for "Management Sins—The Book" for so long and so adamantly that I finally sat down and wrote it. It was a bestseller immediately—not just in the land of Mozart and kangaroos, but in Germany and all across Europe. This made a lot of people angry. Every month I got emails from English-speaking managers: "Do you expect us to learn German, or what?" No, that would be a sin! And that's what you want to avoid in the future.

I'm sure you, as a manager, have got your stuff together and are completely competent in your field and have already achieved a lot (otherwise you wouldn't be where you are today). When something goes wrong from time to time, you probably ask yourself "Why does this always happen to me?" We usually notice, with great pain, when we commit one of these little management sins. We recognize how damaging it is for our success, reputation, career, satisfaction and also the effect it has on our personal relationships.

These, in effect, *not* so little sins, are absolutely avoidable. And in the following pages you can find out how.

Work Yourself to Death!

"Everything great in the world came into being because someone did more than they had to."

Hermann Gmeiner, Founder of SOS Kinderdörfer

Every (Good) Manager Works too Much

If it's a sin to work too much, then it's a sin that we're all committing together. But is it really a sin? "I feel like I have to work 24/7 just to get my regular work out of the way." I hear this a lot from most (good) managers. But what's wrong with that? At the end of the day it's our job to work (too much). That's what we're paid for. So why should it be considered bad?

It's not a sin to work too much—if it's the "right" work.

Let me start off with a "bad" example. My own "sin". I had just been promoted to divisional head and I was excited. Euphoric even. The motivation was palpable. In my acceptance speech I said what probably every leader says to their employees at this point. "My door is always open!"

The first few weeks in my new position were filled with hard work, but I kept my promise. If someone suggested a meeting, I fought to make the time. It didn't take long before my calendar was bursting at the seams. What's more, I played agony aunt to every wearied soul that crawled into my office with a look of suffering on their face. You can probably imagine how I felt after a few weeks (and at what time I managed to do my "own" work - after normal working hours were over for everybody else, of course). Have you heard the one about the overworked manager? "A manager is standing in front of his door one evening when his little daughter opens it and says, 'Thanks, but we don't want any.'" I can assure you, I didn't think that joke was particularly amusing back then. Because I was the guy standing in front of the door in the evening. But I was lucky. I had my own, personal "ah ha" moment.

Are You Doing the Work of Your Employees?

Late one afternoon I stood in front of my office window, still brooding over a problem that one of my team members had related to me about half an hour previously. Restlessly, turning things over and over again in my mind, I stared out the window, not really noticing what was going on in the world outside. I didn't really think about the busy crowd of people on the street below until it hit me like a bolt out of the blue.

Some of those people below were my employees. They suddenly came into focus, pouring out of the building, laughing and swapping jokes. As they happily returned to their after-work lives, I was standing up here dancing alone with a problem that they had shoved into my arms before leaving. *Nota bene,* their problem ...

> The kids were on their way home from school but the teacher had to stay late?

The boss spends his day doing the work of his employees, then stays late to do his own work at night? Another important question struck me: If I'm doing the work of my employees, what are they doing with their time? Leaving, obviously, at 5:00pm on the dot. You could have knocked me over with a feather.

> It's a deadly sin if you work too much because you are doing the work of others!

The worst thing for me was how disappointed I felt. Up until then I had thought of us as a team. Now I could just imagine how my employees were thinking: "Man, I've still got to get this done! I haven't got time for it, I'll delegate it to the boss. He'll know how to get it done fast!"

I was pretty glad that it was late and no one else was in the building. I didn't want anyone to be on the receiving end of my suddenly burning anger. I really felt like giving one of my dear, sweet colleagues (forget collegiality!) a big piece of my mind. By the next day, however, I had managed to calm down enough to consider my own management behavior.

My Own Catalog of Sins

With each passing minute that I thought things over, the clearer it became to me. What had happened the evening before was not just a one-off, it was a mistake that had practically become a habit. When I made a list of all the things I had done over the day, it was a revelation. Do you recognize any of these behaviors?

- I solve a number of problems that my employees should be taking care of.
- I check things that they could or should be checking themselves.
- I give the things that others have already checked a quick going-over.
- I organize processes that my employees should organize.
- I help some of my employees with their personal business, which really isn't my business at all.
- I work on tasks that others would definitely be able to do—sometimes even better than me!

I imagined with horror the reaction of my own boss, should he ever find out (and my old boss found out everything) what I was doing all day: "What's he doing? I'm not paying him for that! That's the job of his employees!" I began to feel very self-critical until perplexity took its place: Was it perhaps my own fault that I voluntarily did overtime yesterday evening? The following realization suddenly dawned on me:

Managers don't (always) commit sins per se. They are seduced into sin.

I hadn't chosen to do overtime the previous evening, my employees had "seduced" me with a great deal of cunning. How did they manage that?

Don't Get Seduced by Your Employees!

Everyone has, or knows, (at least) one employee who appears to be helpless when it comes to certain tasks. You explain it to them (10 minutes sacrificed) and then end up doing the task mostly on your own (an hour wasted). Afterwards, you find out the employee wasn't so helpless after all. They were just pretending! They could have done the work fine on their own.

My employees didn't present me with a problem that previous day without some ulterior motive: it was shortly before the end of the working day. The problem needed to be solved, but they had more important things to do. Namely, going home. So they led me into temptation—and in a way that wasn't even that clever.

I was once responsible for an IT manager who was much savvier than me in this respect. Whenever I gave him a task, it didn't just come back like a boomerang, no, he expected a lot more than that. I admire him to this day for his talent in leading me around by the nose, getting me to do all of his work. Have you ever had one, or more, of these schmucks in your own ranks? Then you should get clear on the following problem:

The more you give in to the seemingly helpless employees, the more inefficiently they will work in the future (because they have unlearned it) and the more over-worked you will be!

It's like my Grandpa used to say: "It's a slippery slope once you've gotten onto the wrong tracks". And what's even more worrying:

An employee's ability to seduce their boss is usually stronger than the ability of the boss to withstand the temptation.

Have you already noticed this yourself? That you fall for it almost every time? This is exactly what I'd like to protect you from. Let's strengthen your immune system against the temptations and seductions of manipulative employees.

The "Huh?" Question

In the midst of the reorganization of a mid-sized company it occurred to me that one department was always carrying out the required tasks much more slowly and halfheartedly than a "sister" department. I had a word with the department head who revealed himself, within a matter of seconds, to be a huge sinner. He complained: "It doesn't matter if I give them the best instructions in the world—my employees always bombard me with questions: 'What do you mean, exactly?' or 'How am I supposed to take that?' They just don't think for themselves." "No", I said, "Your employees are playing dumb and taking it easy while you take the time to answer their questions. They have figured out that they can gain a bit of time for themselves that way." "How do you know that?" asked the Department Head. "Because it's such an old game! I call it the "Huh?" game. As long as the employees play dumb and ask "Huh?" questions, they don't actually have to do anything. Better still: while they're taking a break, their boss is doing all the work!"

There are managers who have let themselves be made fools of in this way for decades—and not just from their own employees. Many even walk willingly into the trap, making idiots of themselves with hyper-bureaucratic procedures. What's that you say? You're not going to go along with it anymore? I congratulate you on your determination. How can you manage to create better working conditions? By ...

1. ... optimizing (not bureaucratizing) the organizational structure,

2. ... pursuing the right strategies, and

3. ... declaring yourself not responsible for everything.

Optimize Your Organizational Structure!

It's not just your employees tempting you to stay in your office after hours. It's often poor organization that makes you put in a lot of overtime.

A good example of this can be found in the sales department that I was hired to optimize a short while ago. The department's sales volume was too low, although most of the managers were working themselves to death. To the company director it seemed like incompetence. I guessed, on the other hand, that the weak spots had to do with the organization, so I asked the leading employees what they did all day. One of them said, "I write reports! I don't manage to get anything else done, let alone doing acquisitions." The others answered "I read reports! I don't manage to get anything else done, let alone doing acquisitions."

When the director heard about it, he made quick work of it: He simply did away with three quarters of the reports. When the sales figures were still floundering six months later, I asked the sales managers the same questions again, though I could already imagine what they would say. This time they all answered, "How can we improve sales? We're not getting any information anymore!"

It's not *you* who should be working for the organization—the organization should be working for *you*!

If you have to read or write too many reports in order to work properly, then you should reorganize your reporting system. If you have too few (good) reports to reach your performance targets, that's another good reason to reorganize, but more on that in chapter 4.

Don't become a slave to bureaucracy and organization!

Some organizational structures are like weeds: When your back is turned, they grow out of control, eating up a lot of your time. Grab your hedge trimmers and some weed killer every day, but don't check things to death!

Don't Check Things to Death!

Sometimes, when I'm visiting managers, I ask them for fun, after exchanging pleasantries: "What were you doing just before I came in?" I now know what kind of answer to expect: most were just reading some report or looking over a list, such as that of a director at a large financial service provider:

He had actually studied a list of all customers currently in arrears just a few minutes before our appointment—each one of the customers had defaulted on a loan. It must have been hundreds, including all the customers who were late by just a day. Why had he not asked his assistant to "Keep the small stuff off my desk. I'm only interested in the ones who are in arrears for over 30 days—please sort the list that way first!" Why had he not asked this? Because he hadn't thought of it.

Only check the things that actually need to be checked by a manager!

Are you a bookkeeper or a manager? The time it takes to check doodly squat emails is no longer there for doing the really important tasks. That's not good. Don't you wish you could spend more time taking care of the things that really matter? If so, then you have my congratulations again, this desire is the first step in your recovery from "checkitallitus", which is wasting away a significant amount of your precious time. In my experience, it takes a few weeks to free yourself of this "checking" addiction. However, it'll be a lot easier if you remind yourself that you can deal with the essentials much more intensively with the time you've saved.

The Meetings Tourist

I know managers who set up a meeting in response to the simplest request, instead of just answering the question or dealing with the problem with a call or an email—and then they wonder why they don't get home before 8:00 at night. Other managers want to be present in every single meeting that takes place in their department. Do you see yourself reflected in this behavior? Why do you do it when you know you have so little time and are so over-worked? Wait, don't tell me. It's because you don't want to miss out on anything. Something important might happen! You've heard of "minutes", right? If people don't know how to take minutes properly in your company, why not reform them yourself, or switch to results style minutes and train your minute-taker accordingly? Or maybe you're worried that they'll make the wrong decisions in a meeting if you don't take the reins? Who's preventing you from overturning a bad decision later? After all, you're the manager!

Decide what meetings are unnecessary in the next few weeks. Which ones do you only need to be at for a short time? Which ones can you miss entirely? In short, which ones will you strike from your calendar altogether?

A company director planned her meetings as follows: "I only go to meetings with a strategic focus. Or when it's about investment amounts of over 100,000 dollars." Since then she has about an hour of time more each day—for the really important things. Could you benefit from that as well?

Pursue the Right Strategies!

Managers in small and mid-sized companies often work especially long and hard. Why? Because they're so ambitious? Because they have to compete with the big boys? That's partly true. But it's mostly because they're betting on the wrong strategy.

In order to take on the big boys, you have to believe in this one guiding principle: Small is beautiful! And by this I mean: Since we are small we can be flexible—at the end of the day that means more customer-oriented! Isn't that great? Not necessarily—because those who say "yes" to every customer's wish will soon become overworked—and will overwork the entire organization, because they will take orders that they can't fulfill or only with a great deal of effort and no profit—because the company isn't set up to take on such orders, the processes haven't been created, because the employees don't have the necessary skills and so on and so forth. Often enough, the managers have to solve these problems themselves, requiring many, many hours of overtime. The result is managers working like crazy, not coming home before ten at night, distancing themselves from their customers, and finally getting their divorce papers delivered directly to their office door.

If your (implicit) strategy means working yourself to death, why don't you change it?

These types of strategies can be changed with very little effort and often without a lot of bureaucracy. The CEO of a mid-sized company in Bavaria once said to his sales manager: "We're not taking any more orders that result in me having to stay until 10 o'clock every evening to iron out the problems that our office workers created. We have to take orders that our people can manage without my help and still reach our sales targets." The sales manager knew exactly what customer orders his boss was referring to.

Declare Yourself Not Responsible for Everything!

Have you ever heard a manager say:

"Sorry, not interested."

Then you would hear a manager who had time for the important things. I know a lot of people who will work out a figure to its third decimal place, who check over every single cent and then complain that they can't go home in the evenings and have to leave their "real work" sitting on their desk!

Are you spending time on the nickel and dime amounts? Or on the millions?

Then start saying: "Sorry, not interested" more often, when someone tries to waste your time with trivialities. Are you worried that these people will be horribly offended? Think you're a heartless manager? Stop liking you? Think again!

A manager who can honestly and clearly say what's interesting and what's not, enjoys not less, but more recognition and respect.

Conversely, a boss to whom you can delegate "any old shit" (direct quote from a colleague), will definitely not be respected by her/his employees. Don't give in to the temptation that you have to take care of everything. For example, by saying to yourself repeatedly:

- I don't do work that my employees could do.

- Before I solve a problem, I ask my employees first "How would you solve this?" He or she should think for themselves, I'm not doing it for them!

- From now on I recognize when someone is trying to delegate their work back up to me. I shoot down this attempt immediately!

- I've lost my passion for detail and I'm breaking the habit of perfectionism. Perfectionism doesn't pay! It takes too much time to work out numbers down to the last decimal point and in the end results in nothing (at least not for the manager, only for the office staff).

- I'm saying goodbye to a culture of mistrust. Checkitallitus takes up too much time.

First and foremost, break your employees of the habit of bothering you with every little question that comes up and asking their: "Mommy, daddy, how does this work?" Set clear boundaries instead and remember: "I ask the questions here." And ask your employees, "How would you solve the problem?" When your employees, after years of disempowerment and forced helplessness, react with dumbfounded looks on their faces, explain to them:

> "From now our roles have changed: I'll ask the questions, you make the suggestions, and I make the decisions!"

You are not the sweeper-upper in your team. You are the one who makes the decisions. If you act like the mailroom guy and do the legwork for your employees, then you'll soon have the reputation of one.

Make Your Sins Count!

No, I'm not saying you have to let go of your habit of working until ten at night. Among managers, the saying still holds true: The highest one on the totem pole is the one who works the longest!

Work as long as you want. But take care of the right things. The important things!

That means: Use your time to develop the right strategies and give high flying new projects their impetus—this gives your area of management, and your career, a decisive edge—but not for pointless meetings, useless reports or to get the work done that your employees are paid for! That is the real sin—against your good reputation, your career, your success, your family and your health.

Karōshi, the American Way

During my time as a manager a colleague once asked me:

> "Don't you check all the outgoing payments?"
>
> "Are you crazy? That's hundreds per week! That's what I have systems for!"
>
> "But these need to be checked by someone!"
>
> "Do you really think I'm being paid to do the work of a bookkeeper?"

My colleague really believed that. He was paid a management salary, but in his heart he was a bookkeeper:

> "But then the guys in the paying office can really pull the wool over your eyes ... !"
>
> "If that were true, I would have hired the wrong guys!

> That's what interviews and performance appraisals are for—to separate the wheat from the chaff. If I employ the wrong people, I can't make it better by checking everything over and over again!"

So why does my colleague continue to waste time? Because he can't overcome his addiction to checking. His employees, on the other hand, have an easy time ensnaring their control freak boss: they only have to dangle a twenty page list of numbers in front of his face and he's off and running, hoping to find the tiniest decimal point mistake. The employees know the weaknesses of their line managers and they take advantage of them (consciously or not)! Do you want to let yourself be tricked in this way? Don't play along anymore! Stay alert, keep your eyes open and watch out for temptation.

You can only resist temptation if you're aware of it.

Best Practice

Some managers may be consistent and not give in to the temptations of their employees. However, they just can't seem to find the right words to explain what it's all about: "Don't waste my time! That's not my problem!" That may be well meant, but with this kind of rebuke you're throwing the baby out with the bath water and your employee is now totally demotivated.

Best practice managers do it like this: "That's an interesting problem that you've just explained to me. Tell me, how would you approach it? Really? That's great! What about ... ? No? Have you got another idea? Good, then go ahead and do it just like that!" The employee will do it too—and with a lot of motivation, and without wasting the line manager's time.

I know a director who doesn't show up to work until ten in the morning and then leaves again at two o'clock in the afternoon. It's been like this for years. Why did no one grumble about it? Because she landed two big new customers every month, which secured the existence of the company for years to come. Why should she stay late checking travelling expenses? That wouldn't have brought her company any further forward. That's what she had her administrative manager for.

If you concentrate on the important things, you give your company—and your career—an edge.

Believe me: A director who checks travelling expenses worth $17.50 will enjoy a lot less success and appreciation than the one who brings in the big fat deals. And what a director does, doesn't remain secret for long. Within three days even the student in the mailroom has heard about it.

A particularly good example of this was brought to my attention by an employee in marketing when I asked her:

> "When did you last see your boss?"
>
> "Oh, hmmm, that must be a few weeks ago."
>
> "How do you know what to do, if she's not there to tell you?"
>
> "You're kidding, right? I know my job. I don't expect my boss to sweat over the small stuff. I only go to her with the really big things."

A colleague from the assembly line who was standing next to her said:

> "You're lucky. My boss is breathing down my neck every ten minutes ..."
>
> "Doesn't he have anything better to do with his time?"

No, he's wasting it and he's doing himself and the company a great disservice. He's a guard dog, not a manager—although he's bringing in a manager's pay check.

Take the Sin Quiz!

It is difficult not to sin as a manager. Temptation is lurking around every corner. That's why I'm always asked if there isn't some trick for dodging everyday time-wasters more effectively. And there is:

> You wash your hands regularly, you should also regularly ask yourself: "What I am doing right now and why?"

The following quiz will help you avoid sinning:

- How much time am I spending on this task?
- Is this task one that corresponds to my position?
- Could someone else do it?
- Does it give my company an edge if I'm the one who does it?
- If I don't think my employees are ready to take it on themselves, what could I do to prepare them for this task in the future?

Not long ago, a divisional manager helped a temp in the imports department look up a zip code in Brazil. A tricky search problem. What a nice guy to offer his help. But did he really help his company get ahead? Or his own career? Not in the slightest. In fact, he was doing them all a disservice: his employees, because he enabled their helplessness; himself, because he wasted his precious time on the small stuff, and his company, because he dealt with something operational instead of focusing on the strategic. His reputation today? No one has forgotten the *faux pas*, saying things like: "He earns five hundred thousand a year and spends his time looking up zip codes! Hey, I'll do that for half the salary!"

The Deadly Sins

One of my old bosses had worked the entire year pretty hard—very little free time and lots of business dinners. Feeling a little guilty, he went to his annual check-up. He got up on the ergometer, started to jog a little and then dropped off dead, just like that. The doctor's comment: "The ergometer is the most frequent cause of heart-attack I know of."

Karōshi types don't stop sinning till they fall down dead.

We all sin, no question about that. You and me both. And even after reading this chapter, you won't be able to break the habit completely. And that's ok.

What's a lot worse than sinning is gritting your teeth and carrying on anyway—even though you know, deep down, there's got to be a better way!

What's really bad is taking pills or drinking alcohol to make the sinful work go down better.

Don't wait until you're in desperate need of a vacation!

Take a break and get some rest. But: don't tell yourself "it'll pass". Make a doctor's appointment at the first sign of exhaustion immediately—do it for yourself—and then go to the appointment come hell or high water. It would be much better, of course, to do it before the first signs appear.

Why Do Managers Sin?

Shouldn't we have answered this question right at the start? No, because this answer goes right to the heart of the problem.

Managers don't work too much because they are 'movers and shakers' who like to get things done. They work too much because they're afraid.

You're not going to hear many who'll admit this, and if they do, then only after a couple of beers. I once watched a client check his calendar with a look of desperation on his face. He began singing the same old song: "There aren't enough hours in the day for me to get my work done" to which I replied, after a quick glance at his calendar: "Couldn't you have cancelled, postponed or delegated this appointment? That would have given you an extra half hour." Without a moment's hesitation he shot back with: "No way, then my calendar wouldn't be full anymore!" "And if you don't have a full calendar ... ?" I asked him back. What he really thought, he didn't dare to say: "... then I'm not needed here in the eyes of my employees!"

Many managers believe that they haven't earned their salary and title if they aren't overworked.

You know what they say: "Only a stressed manager is a good manager!" I like to tell clients who believe this the following anecdote:

Two managers meet up at the 8th hole. One says: "I've tried about a million times to reach you in the last three days in the office. But you're never there!" The other one answered with a smile of satisfaction: "That's right. And this year we've got another double-digit increase in turnover. How *do* I do it?"

It is completely normal to measure yourself, and your work, in terms of how overworked, stressed and exhausted you are. But healthy, pleasant, admired people have different criteria by which they measure their performance as managers—their success!

If you want success, you don't necessarily have to be the last person to leave the office. Mr. Whole Foods and Ms. Versace don't scan a last minute yogurt or shawl through the cash registers at night. They don't sell any of their products, they sell the brand Whole Foods or Versace.

What is your job anyway? To fall over dead? Or to lead your team to success?

You're not going to get far by playing the salesman, temp or bookkeeper role in your team.

The Chapter at a Glance: Take the Wine Test!

When you pour yourself a glass of wine tonight, take a moment to relax and think over the following questions: which of my tasks today could, or even should, have been done by someone else? Which activities will I delegate, cancel, shorten and/or postpone tomorrow? Which tricks from my employees, who want to delegate work back to me, will I no longer fall for?

Don't Tell Anyone What You're Planning!

"Treat your subordinates the way you'd like to be treated by those above you."

Seneca

Another CEO Bites the Dust

When I met him, he had just been appointed the CEO of a large company. He was a man with a proven track record, with several illustrious positions (and companies) behind him. The proverbial "mover and shaker". A true manager.

His predecessor had made some large expansions to the company, adding new business areas and branches. He had recruited hundreds of new employees. That meant high costs and a huge investment. Now, international investors were saying "Time to get down to business". The investments needed to start paying off, now, with a big fat ROI.

The shareholders had set their targets high: an increase in turnover from 50 to 100 million dollars over the next three years and an increase in profits of 8 million. This seemed more than ambitious in the current market situation, even for the experienced CEO. But he wasn't worried.

For two weeks he locked himself behind closed doors with his three board member colleagues and hammered out a new strategy in which the ambitious goals could be achieved. An old fashioned supplier of specialist goods, with a fixed base of long-standing customers, they now had to focus on cultivating a new market and adopting a new style of acquisitions. Or as the sales manager put it: "For decades our customers have come to us because we are the best in our field. Now we have to go out to new customers and try to win them over. That means: active, aggressive market cultivation, more customer orientation and dynamic acquisitions." Sounds great? You bet!

Three months later the sales figures were disastrous. Practically nothing had changed! Even worse: The competition had gained a slight edge in the market share. The shareholders drew their own conclusions: the new CEO had fallen flat on his face. Why?

Tracking Down the Sin

You have to give the CEO some credit. At least he reacted quickly. After the first quarter results were in, he called me for some help. As a fire fighter.

But now *I* had a problem. I'm not your average "fire-fighting" consultant. When I got the call, I knew I didn't want to solve the problem the way consultants normally do: look at some Excel sheets, pour through lists of statistics, make a PowerPoint presentation and then fire 10 percent of the workforce. I've been in the business for over a decade and that isn't my forte. I guess I can forget about a job at McKinsey's ...

No, my approach is a lot simpler. I need to talk to the people actually doing the work. Therefore, the first management tool I always ask for is a set of wheels. The board members usually ask: "What do you need a car for?" The answer: "So I can talk in person with the managers and employees on site; go out to the factories and the branches." Most board members show surprise: "Well, if you think it'll help, we'll get it arranged."

After a few days out in the field, I didn't need Sherlock Holmes by my side to figure out what was wrong—and I never would have guessed it with emails and phone calls. A person standing in front of you answers a lot differently than they do on the phone or in writing. You don't have to be a consultant to have this kind of profound insight. Or a Nobel prize winner. I've had dozens of talks with dozens of people "on site"—and all of them sounded pretty much the same:

> Me: "Why aren't you making more sales?"
>
> Person on site: "Because our products are too expensive!"
>
> Me: "But your regular customers have always paid these prices."
>
> Person on site: "Yes, the regular customers. To new customers we're too expensive."
>
> Me: "Does the board know this?"
>
> Person on site: "Since when is the board interested in operational business?"
>
> Me: "Since your company has a new strategy which is now targeted towards new customers!"
>
> Person on site: "We have a new strategy? Well that's news to me. Wait a minute, you're right, I think there was something mentioned a few weeks ago. And that's the new strategy? More customers? With these prices? How's that supposed to work?"

This is where the conversation usually starts to become interesting. But I'm not being paid just to have interesting conversations. My job is to tell the CEO why he fell from grace. Can you guess why?

The CEO's Fall From Grace

The board members couldn't believe that I was handing in my report after only a week of research. The managers sat together opposite me, their eyes wide in anticipation. I concentrated my energy into breaking the news gently: "Gentlemen (unfortunately, there were no women present), the problem is not your strategy (audible sounds of relief). It's absolutely on target. There's just one problem: no one outside these doors knows what it is!"

"But that can't be true!" the CEO exclaimed, jumping up excitedly. "Three months ago we threw a huge party for all the division and department managers and announced the strategy! The event cost us a fortune!" I dug a bit deeper: "And how do you explain the fact that some of your salespeople in the local offices, yes, even in the main office, are still spending all of their time taking care of the regular customers while others think the new strategy regards changes for public sector clients, and still others are concentrating on landing big customers and yet another group is trying to go into high volume business?"

I have never before experienced such deafening silence at a board room meeting like that.

What those salespeople were doing we call "Managing like Sinatra" in consultant lingo: "I did it my way ..." Managers just go off and do their own thing. Employees are like chickens with their heads cut off: running this way, now that—no system, no direction, no strategy. And that was also the case in this situation, because the CEO had committed one of the deadly sins in strategic management:

> If you simply announce your strategy, instead of selling it, you're going to fail miserably.

Implementing a strategy isn't achieved by organizing a one-off event, otherwise "strategic management" would be called "event management". Think of it as a mid-term communication challenge, which is not necessarily the responsibility of your managers.

> Managers announce strategies. Leaders make sure they are implemented as well.

The actual management sin was committed much earlier by the CEO:

Involve Your Team Right from the Beginning!

Remember? The CEO went behind closed doors with only three (!) of his top managers to work out the new strategy.

> If the strategy development starts off as a secret, it'll stay a secret in the subsequent implementation phase too!

You're probably thinking: "But I can't invite Bob and Ted to the strategy development meeting as well! It would be chaos!" That's right, and chaos is no fun. It's even worse, though, when a strategy is not implemented or gets blown out of the water. At the end of the day, it's not that difficult to gather the troops around the flag:

> Do you want everyone to support your strategy? Then involve everyone!

Does that mean that you should hand over the creation of strategy to others? To your average employees? Of course not!

> Employees don't want to make the decisions on a new strategy. Most people don't want that much responsibility. But they want to be heard.

I had a boss at the beginning of my career who was great at this. When I had a mental bolt of lightning, he said: "Great idea! That will really help us move forward. Unfortunately, I won't be able to take your advice in this case and I'll explain why: ..." Was I disappointed? Of course not! I usually grew about two inches taller at the idea of the big boss liking one of my ideas! And his reasons for not implementing them were always understandable. That reminds me of another boss. Let's call him the chess player.

The Chess Player

He was an exceptionally gifted strategist, playing chess at tournament level in his free time. As a boss, he was also a bit like a chess player. He never liked to share what he was thinking. None of us ever really knew what he wanted or where he was headed. When he announced a new directive or strategy change—it was always in a memo—can you imagine the reactions?

> "Now what?"
>
> "How did he come up with that idea?"
>
> "I haven't got a clue what he wants—do you?"
>
> "That won't work in a million years. Doesn't he understand the current market situation? He should have come to me—I could have told him that right off the bat!"

One day the chess player invited one of us home for dinner. The next morning we were all in the office early. As the lucky guy came in, we grabbed him and started the interrogation:

> "Come on, let's hear it! What was it like?"
>
> "Well, he's got a beautiful house, beautiful kids, knows a lot about wine—I was impressed!"
>
> "If you want to live through the morning, then cut to the chase! What did he tell you about his plans or his strategy? Did he bring up the markets or our portfolio?"

"Nothing! He just invited me over for dinner, nothing else. We ate—and that was it!"

And that was it for the chess player too. We began officially "managing like Sinatra". Everyone just did what they thought was right.

Openness and understanding, on the other hand, will contribute to your strategy's success.

Managers sin when they develop strategy in secret. And they sin while communicating their strategy—although it's really not that difficult.

Kindergarten Communication

I was once walking along a corridor with a board member when he suddenly stopped dead in his tracks and turned to go in the office of one of his employees who was on the telephone—the poor guy was caught off guard, yet he managed to keep his calm. Unperturbed by the presence of the almighty, he finished his talk with the customer. Afterwards the CEO said to him: "You have a wonderfully engaging style on the telephone and you sold our new product perfectly. But if I ever hear you tell a customer: "That isn't possible. The problem can't be with the product", again, I'm going to ride in here like one of the four horsemen of the Apocalypse. Our strategy is customer orientation! The customer should never feel blamed if he has a problem. We're here to solve those problems." "Oh", said the employee visually taken aback. "I guess I forgot about that."

"He forgot the strategy? How is that possible?" I kidded the CEO. He answered: "Don't laugh. When you've been giving advice to your customers for years, convinced that "The technology is always right!", then it's probably not hard to forget at the end of a long day that the focus has suddenly changed." "So this is why you've got to keep giving the same message to your staff every day, a bit like a kindergarten teacher", I surmised. "That's right. My managers and I employ a kind of kindergarten communication."

> A manager cannot believe that employees will forget even the simplest of things. A good leader won't mind patiently communicating these simple things dozens of times a day. Because he/she knows: that's what it's all about!

Or as Mr. Robbins says:

> Repetition is the mother of all skill.

You didn't learn to ride a bike on the first attempt. And it's a lot more difficult for a company to adopt a new strategy. Furthermore:

> The less you talk about your strategy the more others will talk (incorrectly) about it.

That's how nasty rumors get started that can slam the brakes on strategy implementation. If you don't want these rumors to fill up the communication vacuum at the water cooler, make sure that this vacuum doesn't arise in the first place.

Best Practice: The Boss Holds the Door Open

My first boss was the owner of a consumer electronics business. He was always preaching to his trainees: "The customer pays our salaries. So take care of them accordingly!" None of us really knew what he meant back then—although it's so obvious now! Every good upbringing should include teaching this principle. And it's exactly that, that's missing in strategic thinking: would, could, should—the conjunctive is the language of failure.

I noticed one day how my boss held open the door for a customer carrying two heavy bags. That's when the light went on for me.

You can't expect your employees to develop good behavior if you don't live it yourself. Set a good example!

I held open doors like an Olympic champion after that, wishing customers a good day, smiling, asking politely how they were—and against all odds, our business was able to hold its own in our little city when one day a large electronics company moved into it. Nothing against the big names (I buy there too), but the big guys didn't have store managers on the ground putting their strategies into practice like our boss did. That's why we didn't have anything to fear from Goliath. We beat him strategically. That's the power of strategy!

A strategy that is lived by employees is one of the most powerful tools of success in the world.

Why didn't I think of opening the doors for our customers myself? Because employees seldom come up with something that obvious on their own!

Do you want to wait until your employees think of what strategy they should follow on their own? If that worked, why would companies need managers in the first place?

I am so happy each time a manager realizes that it's one of his or her fundamental tasks to explain the company's noble strategy, over and over again, tirelessly and in detail, and ensure their employees understand it:

> "From now on we are customer oriented!"
>
> "Sure, OK boss, and what does that mean exactly?"
>
> "No problem, I can explain it to you again … "

Customer orientation, market leadership, portfolio expansion, forward leaning—no one understands what's meant by those words on their own!

Why not help with the interpretation? Translate your official company strategy from management speak into employee English. Sound crazy?

It's a rewarding but tiring exercise. So now you need to stretch your legs a bit and unwind: it's called "management by walking around".

Management by Walking Around

Strategies are usually not (or not well enough) implemented. This is no exception, this is the rule. It's frustrating but on the other hand here's one consolation: you're not alone. Many clever people feel the same as you. And what do the others do in this situation? What do you do?

Most managers probably have no idea why their strategy (changes) or directives are not successful. How would you handle it? "Write an e-mail", "Send out a memo", "Call one of the managing employees"—those are the most frequent answers. You wouldn't have answered like that, though, right? Man, you learn fast!

When I get called to help drive forward a strategy which is not being implemented correctly, I do something that all top managers ought to do: walk around. Or drive around. I grab a car, get on the road and go out and talk with the people on site. You can't imagine how happy they are! "Finally someone is talking with us" is the most frequent comment I get. "Why did it take so long?" And we're not talking about the weather or our personal lives. We're talking strategy:

> "Why are you here and what's your job, exactly?"
>
> "Um, what do you mean?"
>
> "Let me try again: Imagine that you're a customer. Why should you choose this company and not go to a competitor?"

"I don't know. Why?"

"Strategy is what gives you the answer to this question. It gives you the reason why customers have chosen your company. And now let's talk about how we can translate this "management speak" strategy into real English and how it translates into a normal working day."

"That's a good idea. Why didn't anyone talk to me about it before?"

Employees often expect a lot from their managers—and (mostly) get very little in return. Often not even the minimum: a few unambiguous words about the strategy.

I know managers who have a guilty conscience about this. And those are the good ones. The less good ones say: "Why? Employees should know what our strategy means. What am I paying them for!" That's no longer a sin, it's blasphemy against the spirit of leadership.

The onus of implementing a strategy does not lie with the workers, it is clearly a management responsibility.

"Now just hold on a minute!" say a lot of high level managers to me at this point. "Why does a director have to sell a strategy? It should be the local divisional and departmental managers explaining the strategy in detail to their employees, including what it means for them in practice!" Bravo! Right! And? Do you know any managers able to do that? Could you?

Many managers can feel overwhelmed when expected to pass on the company strategy to their employees and explain what it means in detail.

How can they? They don't learn this anywhere. So they should get help. For example, with a workshop that gives them practical training. The investment pays off: afterwards the team will take off like a rocket. You still have some questions? I thought you might. Everyone has questions. That's why I've put the most frequent into Q&A form.

The Strategy Q&A

Question: "We're just scraping by, month to month. No one has time to think about the strategy!"

Answer: "Of course not. Who has time for that? The good news is: strategy doesn't take any extra time. You can communicate it in passing, like the line manager who reprimanded his employee after listening in on a phone call. The main thing is that you do it regularly. What's more, it's every good manager's job to bring the strategic and operational together. You've got to focus on both: the numbers and the strategy. That's also true of other things: you work and you take care of your health—both at the same time."

Question: "As a top manager, how can I know if our strategy is being lived at the grass roots level?"

Answer: "By going down to the grass roots! Put on some sunglasses and become an incognito customer of your own business. The CEO of a telecommunications firm I know sits in his call center for a couple of hours every month and listens in on what the customers have to say. He finds out more about the implementation of the company strategy there, than in any executive board meeting. If you lose contact with the grass roots, you lose your business."

Question: "If I hear through the grapevine that there are problems with strategy implementation, shouldn't I create some new guidelines?"

Answer: "No! A guideline is just words on paper and it won't mind if it's ignored. Look for the source of the problem, then show your employees how they should be working. Then let them try it, and stay until they do it right. That is guaranteed to work."

Question: "Do I have to set an example in everything that I want employees to do?"

Answer: "If your employees say "The boss doesn't care about that!" about one single aspect of your strategy, your strategy is done for. Or have you ever seen a Hugo Boss shop owner running around in Armani jeans?"

Question: "How often do I need to sell my new strategy?"

Answer: "Until it's sold. Don't like my answer? What did you expect? Managers have a big problem with selling, even when it comes down to something cool, like the amount of rips in a colleague's designer jeans."

How Cool are the Jeans?

Recently, I was in a company that had a very distinctive problem (which I was not called in to handle): The dress code was suffering. It was a financial services provider for an upscale clientele. Suits, ties, blouses and dark, knee-length skirts were implicitly required. Then a new employee came to work wearing a lumberjack shirt, then another wearing ripped jeans. The colleagues made snide remarks: "Where's his chain saw?" or "Going to the disco later?" The boss stood aside, dressed immaculately himself, and gave a pained laugh—saying nothing! As more and more employees flouted the dress code, he wrote out some departmental instructions and sent off an email. Guess how well that worked?

"You've got to be kidding!" one of them said. The problem got more and more out of hand. Who reads policy emails? And who takes them seriously? It's much more arresting to get a hand on the shoulder, a look in the eyes and hear the words, "My dear Mr./Ms. Mayer, I didn't want to have to bring this up, but… "

If managers don't have the courage to broach simple issues like dress codes, because they don't want to be seen as uncool, then it's pretty easy to understand why they don't manage to sell their own strategies.

A manager keeps his mouth shut about the dress code! You really have to let that sink in for a minute. With that kind of annual salary! And when you consider that the same manager is never short for words when a meeting has gone way over time.

Eventually, a department manager from the aforementioned company took pity on his boss and went into the lion's den in his place. One by one, she took her poorly dressed colleagues aside and explained to them in a friendly yet assertive way, why such clothing was unacceptable. She told them what had led to the dress code in the first place, what effect the dress code had on customer relations and how the employees should dress in the future. Was there a huge revolt afterwards? No. The employees had just been waiting for someone to finally draw the line. Did they think of the department manager as uncool after that? Quite the contrary. The employees in question said, "Finally, someone explained the policy to us in an understandable way. Why didn't they do it sooner? Why didn't the director tell us himself?"

Employees don't want to be managed, they want to be led. Do them this favor!

And if it feels a bit awkward to talk about ripped jeans, see-through blouses or any of the other strategic details: start small. Practice your texts in front of a mirror at home, if it helps. Take baby steps. Get some coaching. But, for God's sake, do something! Who else is going to communicate your strategy?

A Leader and Strategist Rolled into One

How do you react when you hear or read about another company's strategy? I normally shake my head. Because I've learned that the reality is usually a lot different. When I heard about a company who pitched themselves as "the best technology with the best service", I laughed out loud. But I stopped laughing when I got a call from that company a few weeks later.

When I arrived a man, who was clearly an engineer, was helping the doorman to arrange the lighting more advantageously for some paintings on display. Both were talking about it as if it were a science. Fascinated, I stood and watched. As they walked across the courtyard into the next building, I witnessed the engineer opening a parcel he had obviously spent a long time packing without muttering a word of complaint. The parcel couldn't be loaded from the ramp because a customer's driver needed to check one of the details. A feeling of dread crept over me, slowly.

Here was a company, which to all intents and purposes, was implementing their strategy on a 1:1 basis. What did they need me for? But more importantly: How had they managed it? When I got to know the CEO, the mystery became clear at once. The man didn't just live the strategy, he embodied it! While I was waiting for my appointment, I heard him through his open office door as he said to his R&D manager: "I don't care if that order pays off or not. If that customer needs it then dozens of others will too. So we have to develop it. It'll pay for itself by the third order at the latest. Even if they have to work overtime, we'll find a solution!" To which the outraged R&D manager said, "Why more overtime? It's our job to create innovations that also work in practice!"

I asked the CEO how he could afford to spend all his time "living" his strategy. Didn't he have other things to do? "Naw", he said. "The entire administrative crap is done by my secretary, my assistant and the individual departments. That's what they're paid for. I'm the boss. I spend my time almost exclusively on the things that keep our company alive: quality and service." "Isn't it exhausting for you to live your strategy 24 hours a day?" The CEO laughed:

"If you're fully convinced of your strategy, it gives you amazing power and endurance—you'll have the time of your life and success. A good strategy is the best motivation of all. For your employees as well."

You probably know the story of the three stonemasons at a medieval construction site who are all doing the same work. When asked, afterwards, what they were doing, they gave their answers as follows:

> Stonemason1: "What I was told to do."
>
> Stonemason 2: "Cutting stones."
>
> Stonemason 3: "I'm building the biggest cathedral you've ever seen!"

Which stonemason was the most motivated? Who did you hardly have to give instruction to, correct, manage or supervise? Who had internalized the architect's strategy?

Employees are the most motivated when they can identify with what they are doing and have internalized the purpose of their work.

Here is a small excerpt of an interview with the manager of the successful German national soccer team given to a newspaper. "You have to involve the players, not just say: 'Run from A to B!' Instead you have to explain exactly what happens when they run from A to B. Soon you don't have to tell the players anymore, they've gotten the idea and run there on their own. You know it really works when the players come to you and ask: OK, got that, what else can we do?"

Cheerleaders are out. Strategies are in.

You can save yourself the expense of offering a lot of incentives and spare your employees the pain of listening to a bunch of motivational hogwash, if your people understand what they are doing and why. By the way: the aforementioned company didn't need me for their strategy. I was there to give their salespeople a competitive edge ...

Strategy isn't a Marketing Gimmick

Unfortunately, I experience this time and time again: The board members formulate some promising new strategy slogans, the marketing department makes the buttons, pens and glossy pamphlets for the employees and the managers say: "The strategy has been communicated!" But that's not enough.

Anyone wanting their strategy to make a difference in the next quarter's figures has to talk and talk and talk. Daily and consistently. To the employees and with the employees until the strategy has become ingrained into their daily processes, behaviors, measures and goals. Afterwards, it's necessary to supervise, diligently, to ensure that employee behavior is really aligned with the new strategy.

I wish there were another way. But there isn't. But I can promise you this: if you go down this path, you'll enjoy strategic success. Guaranteed!

The Chapter at a Glance: Take the Jogging Test!

When you go jogging, walking, or cycling today, ask yourself: how often have I checked whether my people have implemented our strategy (correctly)? How often did I clear up strategic misunderstandings? How much effort do I want to invest in this tomorrow?

Hire the Wrong People!

Boss: "Who hired this jerk?"

Assistant: "Um, I think you did, sir."

An Embarrassing Mistake

A German sales manager of an appliance manufacturer once told me the following story: The incident had taken place a long time ago, but his blood pressure still shot up every time he talked about it: "We had just hired some new customer service personnel in our office. One day an old man came through the door, stood, waited, and when no one greeted him, began looking embarrassed. For about ten minutes no one paid an iota of attention to him although he was becoming quite red in the face. He was one of our oldest customers and when he had come to us in the past, he was always greeted immediately with a friendly handshake, offered coffee and asked about his family. On this day, however, no one took any notice of him because none of the old office customer service reps were there anymore."

The sales manager shook his head, not wanting to remember that such a mistake could have taken place. He took a sip from his large mug and continued on with his story: "It was a secretary who finally saved the day. She finally went over to the customer, who was worth about a half a million a year to the company, and began talking with him. She didn't know the man personally, but had figured that since his customer number was so low, he must be a very old customer. Now I ask myself, why didn't that occur to any of the new customer advisors who earned at least three times that of the secretary!?"

A good question. But not the most important one. The sales manager had received ambitious sales and turnover targets from the board. He was aiming for the job of European sales manager within two years. His bonus and his promotion depended on achieving his goals and he wouldn't be able to do it without the help of his employees: "It makes me sick just thinking about how those idiots in customer care treat our customers. I mean, I ask you, how are we supposed to make our sales targets with such half-wits? They're playing around at the cost of my career! How many customers have they already offended without me noticing? How many millions have we lost because these morons don't know how to generate turnover that is standing, ignored, in the doorway?" And then he finally asked the crucial question: "Why did I hire these half-wits?"

The sales manager was old school. He loved his job. He stood behind his company and the products and he felt a strong commitment. He knew that he couldn't sell all the appliances himself. He needed his employees for this. Good employees, outstanding employees—and then he hired "such half-wits". What had gone wrong?

Are there worse management sins then hiring the wrong people? Unfortunately, yes!

What's Even Worse

Before you hang your head in shame with a "mea culpa" forming on your lips ('cause you know you've made these mistakes in the past—or are still making them now): Stop! Making mistakes in personnel decisions happens to the best of us. All of the time. Even to me. There's nothing you can do about it. So just stop. You can't hit the bull's eye every time. If you can't get over it, you'll fall into an even worse "sinful" trap and go from the frying pan into the fire:

> It's a mistake to hire the wrong people. It's an even bigger mistake not to get rid of them as quickly as possible.

Before your political correctness barometer goes through the roof: Getting rid of, doesn't mean firing (or killing!) it means, doing things right and first trying to save what can be saved—for example, with a motivational talk (maybe she/he can and just doesn't want to) or a critical performance appraisal (a warning shot across the bow). If that doesn't work, written warnings and termination should follow. I, myself, don't just fire the employees in these cases. I make them an open and honest proposal to change careers—within the company, if possible. Nine out of ten employees will not be angry with you when this happens. Quite the opposite. They've usually noticed that they didn't fit in with the team or fit to the work and they will be just as happy that their martyrdom has come to an end in a fair and appreciative way. At least that's been my experience. But why do so few managers have this experience?

Because a lot of them simply don't feel comfortable having that kind of talk or can't admit that they bet on the wrong horse. They aren't able to say that they made a mistake, chalked one up to experience and will get it right the next time. You won't encounter such adolescent fears in Best Practice. A large French construction company, for example, went as far as to set up a department for the low performers where they sent all their "mistakes". The poor performers were given a desk, but no telephone, pc or work. Everything was set up to comply with the strict employment laws there. Not even the famous and ready to fight French worker's unions had anything negative to say about this work practice. Eventually, every displaced employee would got the hint and started looking for another job. Do you consider these approaches extreme? I think so too, to be honest. Luckily, there are alternatives:

It isn't that important *how* you deal with the recruitment mistakes. The main thing is that you deal with them!

If you aren't able to do anything—for whatever reason—with the recruitment mistakes that definitely can't be made to fit with all the will and the training in the world, then you have to say goodbye.

Whether you solve the problem like the French construction company, or find a nicer way, it's your choice. What's important is - the earlier, the better. Because the employees, who don't fit in your organization or their position, are most likely anything but happy and also want to put an end to the suffering.

Who's the Boss, Here?

When I took over the private customer section of a large company, we suddenly came under attack from a competitor. I warned my managers: "Our people are great. The competitors are going to try to lure some of them away, don't let them go!"

Shortly, thereafter, a department manager came into my office unhappy about losing a large number of people: "Houston, we have a problem!" he said. "Houston doesn't have a problem", I answered, "you do! It's your team!" He looked at me, eyes widening. I looked back at him, just as intently. Because at that moment it became clear to both of us why his employees had gotten away and why so many of his new recruits were mistakes:

> Many line managers think: "I have employees—but it's my boss who is responsible for them!"

I hired the wrong people? Doesn't matter, the boss will help me find a new employee. I treat my staff in such a way that they go to the competition for a paltry extra hundred in their pay check? Doesn't matter, the boss'll hire some new people. Most are in denial about their responsibilities which explains why so many line managers complain: "What kind of morons have they hired for us again?" What many line managers don't realize: such complaints are really self-criticism.

> It's not the Human Resources department's job to select employees, it's their job to hire them. The selection process has to be carried out by the line manager himself.

Or put another way:

If you delegate the recruitment of employees (doesn't matter who), you are committing a management sin that will come back to haunt you: you will almost inevitably get the wrong employees.

The choice of employee should not be left to others—not your boss and not your HR department. That's the job of a line manager!

It always pays to make it clear to your line managers that they, alone, are responsible for their employees.

And the recruiting procedures must be set up accordingly: The HR department can take care of the mountain of paperwork, the line manager creates the job specification and determines the terms & conditions (for example, minimum qualification requirements, no hiring of relatives). If that sounds obvious to you, then take a look at what is really going on in companies. I came to the conclusion that approximately 70 percent of all managers have never created a single job specification in their lifetime. When asked, they say: "I just know what an applicant needs to be able to do!" Usually, these are the line managers who complain to me that their employees will never win an award for their brain power. It doesn't have to be that way. A job specification is not rocket science! You can create one in a relatively short period of time.

Don't Expect too Much From the HR Department!

I have nothing against HR. On the contrary, I love them—especially, my wife, who heads up the HR department at an international bank. And I know that good Human Resources managers also get upset about the following deadly sins. That's why:

Never depend on the HR department for your choice of personnel!

A “sinful” case, that affected my own family, also illustrates this impressively: during her pharmacy studies, our daughter Teja worked at pharmaceuticals firm A. After getting her degree, she had the choice of continuing to work for A, who would have loved to keep her, or going to firm B. Now, she already had the contract with B in her pocket. She wanted, however, to keep working for A, so she had a talk with the appropriate recruiting manager there. He, obviously, had a lot going on at the time and possibly a lot more important things to do than hire a newly graduated M.A. In any case, my daughter gave up after several vain attempts to get in touch by phone and, after waiting for days, she signed, like it or not, the contract with B.

The next day her old department head called her from firm A and asked when she would be starting back at work with them. They needed her desperately and had been extremely satisfied with her work. Both women were now very angry with the aforementioned HR department, however, this was not the right target for their anger.

Consider this: “If the farmer is drunk, is the wine at fault?”

From the point of view of my daughter’s old department, Human Resources had been too slow in reacting. But that they were not at fault, when the best applicant was not hired, is quickly apparent. Do you see the problem, here? Exactly: Why did the department manager take so long to call? Was it too much effort to lift the phone from the hook? Was she unable to find the phone? Were the lines blocked? Had the satellite fallen from the sky? Couldn’t she spare the 30 seconds necessary to tell a high potential: “We want you. Make an appointment to come in and sign the contract, it’ll be ready and waiting for you when the HR department has finished with it.”?

If she had taken things into her own hands right from the start, and let HR provide the administrative support (which is what they are actually there for) then she wouldn’t have missed out on her desired candidate. She committed the sin of not taking care of one of her most fundamental tasks. She managed employees responsibly—but how she got them, didn't seem to matter that much to her.

The best applicants are standing at the door, asking, pleading and pounding to get in, calling the HR department three times a day—and what does the boss do? She drives the high potential directly into the arms of the competition, by carelessly neglecting her duties. That's damaging to business. Corporate sabotage from within. Or do you see things differently? If you see a great quarterback, shouldn't you be trying to move heaven and earth to get that guy on your team? You're already doing that? Congratulations!

At this point in my demonstration a client raised an objection with some exasperation in his voice: "But in our company the HR department really *is* responsible for the selection of applicants!" My nerves were slightly on edge when I snapped at him: "Oh yeah? And who's going to suffer daily with the new employee if he or she's the wrong one? Whose performance goals are being endangered here? Yours or the HR department's?" He didn't take offence, looked at me quietly, then picked up the telephone and announced to his HR manager that he would be taking part in the planned interview that afternoon. The HR manager was pleased with the news, because she hated having to make all the decisions alone and then suffer the criticism later for, supposedly, hiring the wrong person.

> Those who take care of employee selection will get the good people. Guaranteed!

You don't have to place the ad in the paper yourself, or do the short-listing—for this (and much more as well) you have a HR department. It is definitely enough if you take care of the essentials. Recruiting is not a mail order business and it's also not Ebay! Good employees can't be ordered online, although that would surely be a profitable business.

Person Beats Paper!

There is another reason why some line managers make poor personnel choices more often than others. They put too much faith in the application. However:

A resume is not the life, it only says something about certain aspects of the life.

Paper is patient. Any halfway intelligent applicant can make themselves look good in their resume. Many pay professional copy writers to create them—so that they make it to every interview. In the interview the actual person behind the great image is not always one and the same.

If you want the best, you have to do more than read. You have to talk.

You don't have the time? Who said it has to take a lot of time?

Whether an employee is suited to the position or not, is plain to see after a few minutes—if you really talk with them.

A key experience, which presented me with the gift of this knowledge, took place on a rainy day—today I am still thankful for that rain. In order to make the interview less formal, I said: "Wow, it's really raining buckets again today!" The applicant answered: "God I know, it's awful. It's been doing nothing but raining the past three days and the weekend doesn't look any better." I almost ignored the reply, putting it down to small talk, but I decided not to let it go when the question suddenly hit me: "Do you really want to have such a pessimist around every day?" I paid attention more closely in the interview and I was right: The man didn't tire of dropping defeatist remarks around the room like little balls of horse dung. What a ray of sunshine. A boost of energy for the whole department—especially during stressful times, when it really counts. When you want the team to keep calm and stay in a good mood.

Electrified by this surprising awareness, I asked the next applicant the same question and she actually replied:

> "I like it when it rains. Because if I don't like it, it's going to rain anyway!"
>
> "Cool answer."

Did you check your watch? Within only ten seconds I had found out eight things about this woman: She was optimistic, quick-witted, outgoing, sociable, communicated easily with line managers, was articulate, and culturally aware—and this would also mean a pleasant conversationalist for customers. This character trait and talent which I had noticed within seconds, was confirmed in the minutes to follow. Her competence was similar to that of the pessimist. Which one do you think I hired? By the way, I would have hired this applicant anyway, even if she were less qualified. Specialist know-how can be learned—optimism can't. *Ergo*:

The most important pro or contra criterion for choosing an applicant is not in the resume! You can only discover the character of your potential employee in a personal talk.

Resumes aren't going to show you a candidate's characteristics, attitude or things like courage, assertiveness, resilience, integrity, teamwork, tolerance, sociability, manners or an ability to think outside the box! These personal traits are a hundred times more important than specialist know-how, but they won't be found in any reference and the HR department won't be able to test them. I have to know and be able to decide who fits in my team. But watch out, you can run into the danger of committing the next management sin, by letting your emotions and feelings of sympathy get the best of you.

Sympathy is Not Your Best Advisor

I provided advice to a client who had gathered together a real army of losers—everyone in the company was aware of the problem. Even the competition knew, and they were leaving him in the dust as a result. The only one who didn't know it was my client. He knew that his team was not achieving its goals, but he didn't know why. Do you want to take a guess?

Weak bosses clone themselves. They hire mini-me's.

My client was athletic ... in everything. So he hired athletic types—without being aware of it. Or as the British say: "People that are like each other, like each other." That's why this athlete had a lot of Arnold Schwarzenegger types in his team (nothing against Arnold Schwarzenegger), but had no innovators such as Albert Einstein, no movers and shakers like Jack Welch, and no one who knew how to inspire like Oprah Winfrey. There were also no planners and no craftsmen.

Watch out: the sympathy factor will mislead you! It's better to pay attention to the character traits and qualities that your team is lacking.

Not long ago I put together a program for a German Business School on the topic of international project management. One of the professors had the following comment: "Your lecture was almost entirely about soft skills. Couldn't you also show us how a person creates a project plan?" "No", I answered. "Tell your students to use Microsoft Project or a project handbook." A fourth grader can read about technical know-how. How many projects become delayed due to a lack of technical knowledge? Exactly. Most problems arise because the team members couldn't understand each other or couldn't cope with the client or contractors. That's why soft skills are so important!

Technical Know-How is Over-rated

This point really came home to me at breakfast in a hotel in Barcelona one morning. One of the waitresses was a real gem. She filled up the bread basket before it got empty, she cleared the tables and greeted every guest pleasantly, even saving some guests a trip to the buffet—at the same time two of her male colleagues were giving a more or less standing ovation performance of lackluster Spanish nonchalance. And now for the best part: the other two gentlemen were service professionals trained at a highly acclaimed hotel trade school. Friendly service, however, is "only" learned by doing. Guess which one got the most tips at the end of the shift. At a tip ratio of ten to one!

Those who depend on technical know-how alone have lost their common sense.

Of course, I don't take my car to a dentist to be repaired. But if I have the choice between a chief mechanic who pokes around listlessly at the engine, and another worker who squeezes himself up to his elbows between the fan belt and engine block, it's a no-brainer. I'm the customer, I don't want a highly-trained and smug "one-trick" pony. I want someone who can get his hands dirty and offers friendly customer service. Of course, the chief mechanic has more technical know-how and will perhaps find the mistake more quickly. Perhaps. But only 10 percent of the customers want first class technical know-how. 90 percent want friendly, customer-oriented service. Unfortunately, hiring criteria is often diametrically opposed to the fulfillment of customer needs: 90 percent of bosses consider technical know-how most important in a potential applicant, only 10 percent deem customer-orientation to be important.

I confess: I'm an Austrian driving a German luxury car. When I say to my favorite mechanic, "Please put on the summer tires." He does this—and without having to exchange a single word, he opens the hood, adds a bit of oil, if necessary (that he doesn't charge me for), checks all the other liquids, the connectors and fan belt. I always think it's amazing—compared with the "service" at other places. Is that due to the technical know-how? I don't think so. That's just good service and customer oriented behavior. The man simply knows what's important.

By the way: I wrote about this issue in a newsletter I created for business customers. Quite a number of readers reacted as follows: "If my boss would pay me more, I would also be more friendly to our customers." To this I can only say: "Then your boss would be an idiot." The friendly waitress in Barcelona was only earning half the amount of the two nitwits! And she was five times as competent and friendly. Performance doesn't have anything to do with money—not for the ones who are really good. And the good ones are the ones that I want. You too?

In consulting sessions, many of my clients are slightly unsettled at this point: "If I'm not supposed to look too closely at technical know-how, what then?" On the things that really count!

What Really Counts

I had an employee in sales who had such amazing technical savvy, he often knew more than the technicians who had designed the appliance—he liked to brag about it every chance he got. The customers loved him. A good man. What's the catch? Can you guess?

Right: The man couldn't sell a thing. This was done by his colleagues. His results were a bit humbling. The less technically competent salespeople waited until the technical whizz kid had finished with his technical oratory which never led to a sale. The colleagues then grabbed the customer and got the deal in the bag.

Employees can be trained in technical know-how. There are plenty of nice seminars and books available. What really matters, however, isn't learned that way.

You can't train someone to have a talent for sales, or have social skills, or an appreciative attitude, endurance, killer instinct, the ability to bounce back, a systematic approach, good self-management, stress resistance ... any other ideas?

You've probably figured out where I'm trying to go with this idea. And if you've gone a step further, that's even better: Now you know how to get the right people in your team. But how can you win over the best for your company? I call it the "plus principle".

The Plus Principle

If you need a customer service employee for your office, what do you look for... stupid question... probably an office worker with experience in customer service. Are you sure about that?

> If you search for what you need, you only get what you searched for and nothing more.

I always want more. I never just look for what I need today, but what I might also need tomorrow—that certain something, the "plus", the "*Je ne sais quoi*". My people should know better than their competitors who only think in terms of the job specification for their vacant position.

As always, I don't get things done at the drawing board, I get them done in practice. I had an applicant who fit my job specification like a glove. She casually mentioned that she worked as a volunteer trainer at her local tennis club. It hit me in a flash "Great, she can make our company known in the entire tennis club!" I hadn't looked for this, I wouldn't even have imagined it, but it could definitely come in handy!

> It makes sense to ask for more than you actually need!

An applicant that only fulfills the job specification—is not enough. These days you need employees that have a lot more to offer. "Yes, I know", said a client to me in answer. "But the big companies get the best ones first. We're left with the bottom of the barrel. The really good ones don't come to us." What occurred to me just then was "if the top talents aren't knocking on your door, you have to go out and get them!"

Go Out and Get High Potentials!

The best ones don't come to you on their own—unless you pay above average or have a reputation like Google.

> Don't wait for the best. Go out and get them!

I know a recruiter at a small company who grabs the best graduates away from the corporations, by picking them up right outside the university doors. He has only a single argument in his favor, "The corporations will pay you more, but you'll only be a small fry there—with us you'll be the big cheese, king of the hill, master of the universe." Who can resist that kind of enticement?

When I have to fill a new position, I don't waste time waiting for the best to come banging down my door. Managers don't wait, they go out and get. So I went out with my bullhorn and made announcements at every corner: "Hey everyone, listen up, I'm looking for ..." I activated my networks, put the job description on LinkedIn and checked all my addresses on file (by now up to eight hundred friends, acquaintances, partners, contacts, etc.) That's how I always get what I want—and even more to boot. A good network is always quicker and better than the official market.

"Yes, but you still have to be able to get the good ones to stay", some of my clients will complain. "If someone is good, they'll be stolen away by the competition! They pay better and they can offer a lot more interesting career opportunities." What a huge misapprehension!

Money Isn't Everything

Just so there are no misunderstandings: Money is important. If you put another hundred in my hand, I will, of course, say thanks. But:

If you can't offer more than a high salary, the people will come but they won't stay.

As soon as one of your competitors offers a higher salary, your employee is out the door. That begs the question: How can you keep good people, although others are offering better conditions? Is that even possible? One look at Best Practice is enough. I have asked the employees of such companies to finish the following sentence:

I'm happy to work here, although I could earn more somewhere else or might have better career chances, because ...

- ... no one interferes with my work and I have the freedom to make big decisions—which is exactly what I want.
- ... my boss is a really great lady who supports me when I've got a problem, whether at work or at home. That's something that money can't buy!
- ... I'm not a small cog in the wheel. I'm responsible for my own division which I can manage the way I think it should be managed.
- ... I can come and go when I please!
- ... we have a super team and have tons of fun.

- ... I get to choose the courses I want to take, and don't have to go to seminars and workshops that my boss, or the HR department, dictate.

- ... in addition to this job I also get to have a life.

- ... they treat you like a human being here, and not like a cost factor.

Impressive, isn't it? These reasons all have nothing to do with money or career. How would your high potentials complete the above sentence? Do you know? The ones who do know enjoy high potential retention.

Only the ones with nothing else to offer have to lure their employees in with cash.

What do you have to offer your high potentials and high performers? It is a management sin to only think about the money. Employees often expect something quite different. The ones who don't take those things into consideration are the ones who have employees that are always looking for a better deal.

What about enthusiasm, great team spirit and fun at work? That satisfies most people more than a few dollars more. If work is not a "must" but instead a meaningful balance of job and free time, you have more than satisfied employees, you have friends for life. Who will go the extra mile to be able to keep working for you. Or as one of these high potentials once said to me, "Don't tell my boss but I would work here for 200 dollars less. The work, the atmosphere, the colleagues, the boss—everything is fantastic. I wouldn't give it up for all the money in the world!" Would any of your employees say that? How can they, if you're the one complaining about your work on a bad day?

Who Motivates You?

You're the role model! How can you expect to have motivated workers who make work fun, if their line manager is moping around not able to say a quick hello? Sound familiar? Sometimes you feel like you've just had enough? What manager doesn't feel that way sometimes?

Motivate your employees as well as yourself!

Who else is going to do it? Most managers expect their bosses to motivate them—or even their employees, customers, colleagues, or the situation. You can forget about that. Motivating the manager is the manager's job. There are five hundred methods of self-motivation. Look for the one that works for you. Effectively and permanently. Cheer yourself on, that's what makes you a leader.

How do I motivate myself? With lots of things. When I am confronted on a bad day with the awful reality of my management responsibilities, I slam on the brakes in a panic and force myself to direct my horrified look to something else—for example, what I want to achieve, what I would most like to have, what would be exactly the right thing at that moment. No, I don't mean a two week vacation in Porta Vallarta, but something that could help me to solve the problems that I am currently facing.

A partner is being difficult. He's angry and doesn't want to talk with me. That's bad. But: it doesn't help to wallow in your sorrows very long. I try to imagine instead how we two could get back on track. For example, swapping a few stories, and having him pat me on the back again with, "Hey, we should do that more often!" My courage in hand I begin to feel the urge to go up to him and break the ice, "Sorry, man, what was it that went wrong? Tell me what you think and let's get it sorted out." And that's exactly what I do. Because I can motivate myself to do that. You can do it too, otherwise you wouldn't be reading this book—because that takes time and energy. Most managers don't do that, because they can't. But you are. So don't try to tell me that you can't motivate yourself ...

Can we Praise our Employees too much?

There are some managers who aren't sure how to praise their employees at work. Most won't admit this before the 2nd or 3rd glass of wine. But every manager knows that the main reason for constant requests for an increase in salary, inner emigration and high potentials quitting, is not the lack of dough, but the lack of the right kind of praise. It's a very real problem. The following checklist can help:

- Praise is not a leadership style. If you praise too much, your praise loses its value.
- Never give general praise à la: "Bob, you're the greatest!" An employee could think, "He has no idea what I actually do, he's just flattering me because he wants something."
- Only give concrete and justified appreciation, and always state the reason, for example: "Amy, you really got order 127 processed quickly. I'm impressed. Such a huge volume in only three days. Respect!"
- The more people around, the stronger the compliment comes across.
- If you don't really mean it, or you can't think of anything to praise at the moment, keep your mouth shut. Nothing drives your employees into the arms of the competition quicker than insincerity.

One of my bosses wandered into my office one day and asked: "How are you doing today?" I am always pleased when someone finally cares enough to ask, so I took a deep breath and began carefully explaining my actual situation. Before I had uttered a single word he was already out the door. He had a long day ahead of him too, and still had lots of offices to get through. I probably don't have to mention that by this time I was definitely not planning to stay there 'til my hair went gray… Which leads us to one of the most important questions regarding employee retention.

What Kind of Boss Would You Like to Have?

How do you get the best employees? How do you keep them in spite of a better paying competitor? Just think about yourself:

> Imagine your ideal boss. Can you imagine being that kind of boss for your employees?

Create a catalog of characteristics that this ideal boss has. Can you imagine exercising these characteristics bit by bit yourself?

> Everyone wants to work with an ideal boss and no one leaves him or her because of a few dollars more.

If you want more security *vis-à-vis* recruiting and retention, add a little something on top:

> Be the boss that you would like to have yourself and, at the same time, the one that your employees would also like to have.

Every employee is different. One wants a lot of praise, the other wants more freedom, a third wants strict management, and the fourth needs a shot across the bow. If you can manage, roughly, to give someone what they need, you will never have personnel or productivity worries. Because:

Employees will go through fire for a boss who knows and considers their needs, and stay with him or her—come hell or high water!

When I asked a veritable genius developer why he still worked for a particular supplier (when he could have earned three times as much with one of the automotive manufacturers who were constantly trying to headhunt him) he answered my uncomprehending look with the following: "I have a house, two cars and three children. Money isn't as important to me as it was 20 years ago. And I don't think that BMW, Daimler or Ford could guarantee that I would have a boss like mine. She gives me the freedom that I need, and can also pull me back to reality when I've gotten a bit lost in my research. She gave me three weeks off work once, when my wife was in the hospital ... and even if I were ungrateful enough to abandon such a great boss—I would have to be totally crazy, I'll never have it this good anywhere else!"

Sounds sensible, doesn't it? What is it they say? First class managers get first class people—and they keep them.

The Chapter at a Glance: Take the Kissing Test!

After you have given your partner a welcome home kiss tonight, ask yourself, what can I do to recognize and hire the "right" employees at work? What did I do today, to keep my best people? What will I do tomorrow?

Do Everything Yourself!

CEO to a consultant: "What do you think is wrong with our company?"

Consultant: "I think I'm looking at him."

The ADHD Manager

Some time ago, I sat across from the CEO of a mid-sized company in his office. We had exchanged pleasantries and were ready to get down to business when we heard a knock on the door. The chief accountant came in. She had discovered that the monthly IRS report was wrong but she couldn't figure out where the mistake was. After both had solved the problem, I turned to the CEO and wanted to explain to him the nature and effects of a new type of customer service. But then his telephone rang—production had a problem with a machine.

After this telephone conversation I managed to finish my presentation when his assistant stood in the doorway and explained that the next appointment had arrived fifteen minutes earlier than agreed. He asked whether he should invite the visitor in. The CEO told him, annoyed, that he should ask the visitor to wait, and then said to me, "We definitely have to discuss this in more detail. Could you come by next week?" "No", I said. The CEO looked at me appalled—top managers are not used to being contradicted. "If you would like to continue our discussion on a more professional basis", I suggested, "why don't you come to my office?" You could have knocked him over with a feather.

Coincidentally, I ran into him the same week at the coffee bar at some event. He made a grimace when he saw me but came over and got right to the point:

"You know, it's always the same. Every day I get interrupted, the same as when you came to visit. Everyone wants to speak with me and thinks that I have the answer to every question! I work fifteen hours a day, six days a week. What do you think is the problem?"

"I think he's standing in front of me."

I actually phrased it a bit more cautiously. I didn't want to give him another shock, after all:

"Almost all good managers who take their jobs seriously, feel the same as you. Almost all. There are a few who have solved this problem rather elegantly."

"Well, then let me have it. What's their secret!"

"It's not a secret. It's a kind of quiz."

"A quiz? OK, I want to take it!"

The Socratic Manager Quiz

"What is your main task as a manager?"

"Making decisions, of course!"

"What kind of decisions?"

"The strategic ones, naturally!"

"Aha, and helping the chief accountant find a mistake in the tax return is a strategic matter, is it?"

"I think I know what you're trying to say."

"About what?"

"That it isn't my job to solve problems at the operational level."

"Yes, but what does that mean?"

"That I have to send my employees away when they ask me operational questions?"

"Then they will be pretty angry. And they'd be right, by the way."

"Because, maybe, when they come to me they are actually bringing along the suggestions for the solution as well. And I only have to say "yes" or "no"."

"That sounds good! That only takes a fraction of the time that would otherwise be stolen from you."

"Hmm, why didn't I think of that?"

Why Didn't I Think of That?

Have you ever asked yourself this question? Then I'll give you a couple of answers to choose from: Managers want to feel needed. They want to feel irreplaceable so they check everything down to the last detail. If a sack of screws falls off the loading ramp, who's going to care? The CEO who knows every particle of dust in the warehouse by name.

The boss who sees, knows and does everything ... better! He's even better than his chief accountant in bookkeeping matters, as our example above showed. When the owner of that company heard this, by the way, she said: "My CEO's doing what? Bookkeeping? Great! From now on he's getting the salary of a bookkeeper!"

It is a very human and fine thing when your gut feeling shouts: Help the accountant! Help the pickers in the warehouse! Do everything yourself! Clean the windows, if the cleaners don't do it as well as you can! Thank your gut feeling for its sympathetic approach—and then switch on your healthy manager common sense.

For example, by asking yourself:

> Why am I receiving a manager's salary? For doing the work of my employees? Or for bringing my company forward, strategically, and making the really important decisions?

What tempts you to do everything yourself? Your management ADHD? Your helper syndrome? Your fire-fighting impulsiveness? Your addiction to checking everything? Your perfectionism? Your mistrust of others doing something that you didn't get started yourself? Whatever it is: ask yourself from time to time whether you really want to give in to these urges. If yes (which is seldom the case) you're in need of a good coach. If not (which is usually the case) you can work at uncovering your individual reasons for these behaviors.

What Tempts You?

Start your investigation at the very beginning—of a normal day at your office. What do you do first?

> "Normally, I go through my correspondence", said the CEO from our example.
>
> "Yourself?"
>
> "Yeah ... why?"
>
> "Who sorts through your correspondence first?"
>
> "Pre-sorting?"
>
> "You have an assistant, don't you?"
>
> "Yes, and a secretary—and both do a great impersonation of the mailman. They dump the mail on my desk and leave."
>
> "Do they know the criteria for sorting your mail, the way you'd like to have it done?"

"But that's obvious!"

"And that's why the mail lands on your desk without being sorted?"

"Do I have to explain every little detail to them?"

"So you think it's better to take care of every little detail yourself?"

We had reached an impasse. What my CEO didn't know was that his company's owner had recently spoken with me. She wanted to know: "Is it time for a new CEO? I have the feeling that the old one isn't driving the company forward the way I had imagined. The competition is getting stronger." Of course, the competition gets stronger when managers would rather sort through piles of mail, help with the bookkeeping and take on other employee tasks as a priority!

Every second that you spend taking care of the work of your employees, you are sawing away at the supports which hold up your job and your division. You are not just damaging your company, you are endangering your employees' jobs. You are doing more damage than any competition or the politics-playing schemers from within the company could ever do.

Or as an amateur skipper and the director of a chemical firm once remarked: "If the skipper's in the hold flipping pancakes who's on the bridge steering the boat through the rocks?" Unfortunately, too many managers are standing at the griddle, which threatens the existence of a company. In many companies, it's not a lack of capital, innovations or a supply chain strategy which has the greatest impact on a company's success. The greatest risk, a huge danger and the worst threat to the existence of a company are the managers who don't make the necessary strategic decisions: wallowing instead, under a landslide of emails in an operational mire.

Hit with the full force of this realization, the aforementioned CEO became white in the face with horror. All this time he had thought his sacrifice of fifteen hours a day was benefitting his company. That he had actually been digging his own, and the company's, grave blew him away completely. He sat down immediately to analyze his normal work day. He was helped by the following questions:

More Questions on Temptation

- Before getting started, do you classify a task, according to "essential/strategic" and "only urgent"?
- Do you consistently delegate everything that is "only urgent"?
- Is it necessary to do what you are currently doing?
- Can you shorten it somehow? Can the problem be reduced to its key elements?
- What would need to be done so that someone else could take care of this task? Training, instruction or coaching for an employee?
- Why aren't your employees more independent? How can you empower them to become more proactive, self-sufficient and entrepreneurial employees?
- Are you taking care of operational tasks? Or are you setting the framework and strategy to enable others to work well?
- Would you rather do everything yourself than explain it for the "hundredth time" to an employee? Why not explain it right the first time so that your employees understand and can do it themselves?
- What are your most important strategic tasks, challenges and priorities? If you could only concentrate on one thing: What tasks would you then have to delegate?

- You can also prioritize: What do you absolutely have to do first to drive the company forward? What else could you do, though it isn't as important? What should you definitely not do anymore yourself?

- How can you be sure that you are spending the least amount of time possible on non-strategic things? What kind of contract could you make with yourself? What checks and balances do you need to introduce? What sanctions? What rewards?

- Who could you bring on board to help you strengthen your self-discipline? A coach? Your best friend? A colleague? Your secretary or assistant?

What's Your Job?

What is sad but, unfortunately, true in many companies, today: Managers prostrate themselves to the dictates of urgency—the really important things are done, if at all, in the little time that's left over. There's a German saying that goes: The fish stinks from the top down. It doesn't mean that managers don't work enough, it's that they don't take care of their real work: the essentials.

The employee thinks about working. The manager thinks about the work. That's what separates him or her from the employee.

This is when the CEO had his "Ah-ha" moment: "As soon as someone comes in with a task for me, I jump in like the proverbial fish into the frying pan. I need to rein myself in more. Because: The employee's job is to finish the work that is delegated. A manager should, instead, be considering who to give the work to—and remembering what the boss's work actually is!"

People need discipline to resist the distractions inherent in busy routines and to concentrate on the essentials. That is not only true for managers, it's true for everyone. I ran into an old friend, recently, whom the doctor had strongly advised to get more exercise. When I saw him in one of his few free moments, he was checking his electricity bill. I must have shown my surprise because he immediately recognized what was going through my mind and admitted with chagrin, "OK, I know you're right (I hadn't said a word). I should be tying the laces of my running shoes right now." Or as the great German poet once said:

> *"What is your greatest responsibility? That's the main task of the day."*
>
> Johann Wolfgang von Goethe

It can also be said in plain English, remember your grandparents' advice: "Think first, act later."

> *"Anyone can solve problems. A smart man thinks about which problems he wants to solve."*
>
> Ernest Hemingway

That is strategic intelligence. That it's often lacking in the required amount at the board level, was demonstrated recently in a large corporation. After the old guy had stepped down, junior took over the executive board and promptly founded a committee for "global equilibrium" (or something like that)—while his sales figures took a nosedive. It's great to save the world, but a company with good sales figures and secure jobs would probably have a better chance at it!

Choose your priorities wisely and strategically. Follow them and ignore the sinful temptations that appear left and right along your path. Do you know what also helps? Revolutionizing the reporting system in your company.

Who Reports to You?

"How many people report directly to you?" I asked the CEO.

"I think there are about seventeen or eighteen employees—if I'm not forgetting anybody."

"If you only talk with each of them for half an hour, how much time do you have left for your other work?"

"Hmm, about an hour—between nine and ten at night."

With all due respect to flat hierarchies—even here, less is more. The ideal, in my opinion, is not to have more than ten employees reporting to you.

Managers usually need time to get used to this first: It isn't necessary for you to see everything and everyone yourself. A great many of your employees should be reporting directly to your "Lieutenants". If there are questions later, you can always go directly to the employee to dig a bit deeper. Don't you trust your management personnel to inform you reliably? Then why are these people still working for you? Train them to report to you correctly—or look for better people.

At this point the CEO became very animated: "But if my people report directly to me, I can make decisions much more quickly, directly, and more flexibly, than if they went through a middle-man first! Our strength as a mid-sized company is that we are fast and flexible. Fast beats slow every time!" Did you catch that error in reasoning? I continued with my questions until he finally came up with the idea himself. He looked a bit deflated, nonetheless: "What I thought was a huge advantage, was actually a disadvantage! For every decision that I made quickly and directly myself, I had four other employees who had to wait for their chance of an audience with me!" And not only that. After a further five minutes he also figured out the next important point. "It's only ever really quick, if the decision is being made where it has the most impact. Employees should not have to waste days and days waiting for an appointment with me before making a decision, they should decide themselves how to proceed."

Empowerment is quicker and better, than doing everything yourself.

Some banks, for example, have already realized and implemented this. Anyone who applied for a loan in the past will probably remember that although they applied to their banking adviser, the application form still wandered its way through the internal post to various checking and decision-making bodies until finally coming back to the adviser via internal post—in this time one could have approved (or rejected) four new applications. That's why modern banks do it differently: Loan approval is made by the one who is responsible for the customer. Guidelines, rating systems and general terms and conditions created by the strategically thinking boss make this possible. *Nota bene*: The boss doesn't make the operational decisions, he or she just sets the basic strategic parameters.

Let your employees work on their own and make decisions on their own—create a framework in which they can do it correctly.

The Manager as Risk Factor

You hear a lot about managers who try to evade their taxes although they receive a shamefully high salary for basically outsourcing jobs abroad and then cutting employee benefits. Even worse than an overpaid manager, is one who checks all their own emails, combs their tax declarations for mistakes, and plays company politics—instead of creating new markets, optimizing the supply chain, driving the process of innovation, increasing profit and upholding corporate responsibility. Or more to the point:

A manager, who's become a bottle neck for the company's success, hasn't earned a horrendous or even moderate salary—he or she hasn't earned one at all.

Or as one board member formulated: "If you pose a risk for the company, you shouldn't get paid for it as well!" A turner, miller or storage worker, who spends their time reading emails, instead of turning, milling or storing, doesn't get paid, they get fired. Unfortunately, I often have the impression that managers try to make themselves indispensable, thinking to themselves, "If my employees are doing all the work, what's left for me to do?" That just demonstrates that many managers have never learned their true calling, to drive the company forward.

Have the Courage to Put Your Feet Up!

A fortune 500 study showed that most companies you and I know of today will be has-beens in 20 years' time. Now we also know the reasons why: If you're rummaging through a mountain of emails, you're killing your chances in the future.

The head of development at a mechanical engineering company is a shining example that it doesn't have to be this way. This anecdote has become part of the learning culture for trainees in all areas of the company: One day the head of development was caught by a director while he was sitting looking dreamily out of his window. His bare feet on the window sill, he was leaning back in his arm chair playing with a yo-yo in his right hand. The director reacted with indignation: "Taking a holiday at your desk I see! I didn't know I was paying you to put your feet up and play with a yoyo. What's wrong, tired of working here?" The head of development answered calmly: "I was just thinking how we could make more money with our products, in light of the saturated western markets and turbulent developments in China, India and Russia, without having to invest millions in country specific development projects." The board member did an about face and quickly said, "and don't you dare take your feet from that window sill until you've solved that problem!"

A manager is primarily a doer. A top manager is someone who can do what it takes to make the ideas, innovations, strategies and clever plans come into being; that helps their company and helps manifest the coveted quantum leap—ensuring their immortality.

Those who can't keep their feet still, may never make it as a top manager. By the way, I don't mean that literally. What I mean is that if I have my best ideas while going out for a walk with my dog, instead of with my feet up at work, the person paying my fee doesn't care. Where I develop the good ideas is my business—I'm being paid for performance, not attendance.

Where do you get the best ideas? Go there and get them flowing!

How often should you go there? Many say an hour a day. Others take a half day per week. Still others are the most creative at the weekend or under the shower. Quite a number of managers go to work early so that they can have an hour of uninterrupted thought before the "rank and file" pour in and the usual creative thought-crushing hustle and bustle enfolds.

If you put your feet up for two hours and develop a strategy in this time for improving production capacities for half a year, who's going to accuse you of wasting time?

Are you worried about having to justify yourself to your line manager or your employees? Do you fear having to explain that you're not wasting time, but rather trying to save your company from going under?

A manager should be brave enough to put their feet up once in a while, get caught, and explain it to the uncomprehending witness.

Management requires courage, endurance and assertiveness. Anyone who believes that they can get by without these traits, is committing a further sin. One of the worst that there is in management. And that's also true for life in general: If you want to do things right, you have to assert yourself enough to stand out from the crowd of amateurs who haven't got a clue.

Best Practice

What goes on in the average workplace, we all know, only too well. Let's break away from the sinful status quo and turn our sights to what we want to achieve: Best Practice.

I know a CEO who is almost never in his office. His secretary rules the roost when he's gone. A business reporter once referred to him as the "opera manager" because he was often spotted at balls, the theater and concerts. He was able to mix with high society much more easily on the golf course and at certain events than in his office. The journalist was just being petty and small-minded, because everyone in the company knew: "If our CEO actually does spend a day in the office, it means trouble. As long as he's hobnobbing with the rich and powerful at cocktail parties and openings, we can be sure that he's setting up big orders and creating the environment for our company to keep doing business." In fact, the CEO has clinched far more deals on the golf course than in his office.

The Braun Parable

Many years ago I spoke with a designer from the Braun corporation, the well-known manufacturer of razors, household appliances and other types of electronic consumer goods. He told me about the company's idiosyncratic yet impressive design process: "When we develop a new razor, we ask our customers, handlers, engineers, designers, managers, marketing people and futurologists what the new razor needs to be able to do. Then we build a prototype and stick everything into it that the people want. Finally, we take another look at the thing and think about how it will be used and get rid of everything that could be a problem." This comparison translates wonderfully into management practice:

List everything that you must, can and should do. Then get rid of everything that doesn't help you move your company forward, strategically!

Does that hurt? Sure it hurts. Who wants to have to say to their chief accountant: "Sally my dear, I love ya, but you're going to have to solve this problem on your own!" That's no fun. But think of the alternative:

What hurts more? If you put an abrupt end to taking care of operational things or if you take a pass on the strategic?

Every working day, matters of urgency want to force us to bend to their will. The operational wants to tempt you to sin. Resist! Fight back! Take care of the essentials—the rest you can delegate.

The Chapter at a Glance: Take the Blanket Test!

When you snuggle in under your blankets tonight, ask yourself:
Did I manage, today, to not do everything (operational) myself, but instead only 90, 80, 70 … 30 percent? And what percent of my work time was spent on the essentials?

Never Sign Anything!

"The management board shunts important decisions to the strategy committee, who then delegate to the divisional bigwigs, who in turn pass it on to a department head—and so on. The most important decisions are probably made by the janitor. Give that man an office at the board room level!"

Angry plant manager at a mechanical engineering company

The Hide and Seek Manager

How long does it take for a decision to be made in your company for an investment of say 5,000 dollars? Think that's funny? A smirk is the most common reaction I get to this question. Usually, followed by sarcastic comments like: "Long live bureaucracy!"

If decision-makers can't make decisions, the company comes to a grinding halt!

Lots of decision-makers haven't really earned the title. They don't decide on something, they wait it out. Do you know such managers? No doubt we can all quickly point the finger at a few guilty of this behavior over the course of a day. But then an uncomfortable awareness starts creeping in. How immune are we ourselves to the sin of risk avoidance? How quick and self-confident are we in making decisions? What personal or business decisions have we been putting off for days? Do we send things back to a committee "for further analysis"? Why do we do this? Why do we postpone instead of deciding? The terrifying answer is: because to some people it makes sense!

Let's consider the example of a divisional manager at a pharmaceutical company. He always sends huge decisions back to the board. Categorically. Whenever his employees put an idea forward which is slightly out of the box, he slams on the brakes: "We can't make an ad hoc decision on this, we need to have it analyzed in detail." His motto is: "Better to decide nothing at all than to make the wrong decision." And for good reason. Past medicine scandals have taken their toll. His cautiousness pays off: he never makes any bad decisions—because he never makes any at all.

Many of his peers are astonished: "Managers being paid for postponing, doing nothing, and not making decisions? That's crazy!" It isn't crazy, it's management. The question you have to ask is: Does it pay off?

Does Avoiding Risk Pay off?

We can answer the question using the following example. The Romanian country manager of a service providing company is known for being cautious. Without a touch of irony he once explained to me: "My strategy is: just don't make any mistakes, otherwise head office finds out and notices that I exist. If I keep my head down, I can pretty much count on my contract being extended." If that's really the case, the only logical conclusion you can come to is: managers, run and hide! Become invisible! The ones avoiding decisions are the ones keeping their jobs. The question is: really?

Suppose the aforementioned manager had the chance to land a big fish for the company—a new customer with impressive contract value. There's only one catch: no cash in advance. If he takes the order and the customer is solvent, he might get a promotion to the Eastern European Country Manager position. If the customer doesn't pay, the strategists at headquarters will notice that he "exists"—and degrade him to branch manager or just chuck him out entirely. Ergo:

Managers weigh things up before every decision. The risk, of making no decision is considered by many to be much lower than the risk of making the wrong decision.

You may, as we do in Europe, make fun of civil servants, but: getting paid to do nothing isn't completely without its merits! After all, it worked for a while for a former German chancellor. His employees affectionately called him "Buddha" and not just because of his rotund figure. He passed on decisions as often as he could with the delight of a child, smiling as he did like his namesake. If this successful chancellor (at least in his own eyes) could do it, then why shouldn't I recommend it too: Managers of the world, put your decisions off forever!? Why not? Because the chancellor didn't get voted in again. Unfortunately, what's true for Buddhas and chancellors is also true for managers:

Up or out! If you wait the decisions out, you might be able to hang on for a short time; however, if you stay in the same position too long, at some point you'll be replaced.

He'll be replaced, because he couldn't drive his division forward decisively, or because there are newcomers already chomping at the bit who aren't afraid of making decisions and can do a better job getting things started. If you are not getting noticed, you're not getting noticed positively. So:

Waiting things out only works for a short time. In the long run, avoiding taking risks is like shooting yourself in the foot.

Oh, you knew that already? But you're even less amused at making bad decisions? Sounds like a real dilemma to me. And how are you planning to get yourself out of it?

The Happy Medium

Postponing a decision is bad, making the wrong decision is no better. If you've got Scylla on the left and Charybdis on the right, the best way is the same as what Odysseus chose: through the middle.

Strike a happy medium: the calculated risk!

Or as the old saying goes: "No risk, no fun!" I had just explained this to the country manager when he looked up at the ceiling of his office and began calculating the risk in his head: "It is definitely risky to go into business with this big, new customer: Will he or won't he pay? But, maybe, if I negotiate a bit, I can get half of the prepayment that we want out of him. Then I only need to build up three or four of our B customers to hedge against the worst case scenario, so that headquarters doesn't notice." Such considerations are the stuff of good managers. Or put another way:

Risk management means determining and evaluating the risks and then using that as a basis for confident decision-making.

You don't need an MBA to do that, just some healthy common sense—and a short check list.

Short Risk Management Check List

Ask yourself before (important) decisions:

- Do I have all the information I need for a decision? Is anything missing? Is the effort required to get the missing information worth it?
- Am I in danger of making a decision based on a gut-feeling alone?

- The other extreme: Am I investing 90 percent effort to get the remaining 10 percent information which will not really effect the final result?
- What do I do if I lack relevant information? Can I approach it with small, consecutive steps as in Igor Ansoffs' "Theory of Weak Signals"?
- Have I gathered, in particular, the information on probable and possible risks? How high is the probability? What damage can be realistically expected?
- Have I done everything necessary to minimize the risks? Or, at the very least, carried out preventative measures or developed alternative plans?
- What is the worst case scenario? And how can I prevent it or compensate for it?
- Have I calculated realistically (!) what this decision will cost? How much will it help, assuming I let go of my wishful thinking?
- What will happen if I decide against it? What am I losing out on?

If you work through this check list, you should only have a small element of risk remaining. My advice, here, is:

> Take calculated risks! You will reap the rewards as a manager for your readiness to assume risks.

However: Although the checklist is easy to use, even for managers without an MBA, many still fall flat on their faces.

Analyzing to Death

When I tell managers: "Manage the Risks!" many hear "Analyze them to death!" For example, the marketing manager of a mid-sized company: when her CEO asked her for the implementation strategy of a new product, she answered: "I'm not able to report back on that yet, the results of our market research are still not available!" The CEO argued that these had been available for months already. "Yes, that's right", said the marketing manager, "but the studies were not conclusive enough. We've set up a new study to verify the first one." What's your reaction on that?

It was plain to see that the marketing manager was trying to avoid making the wrong decision. One could say she was guilty of "paralysis through analysis". The strategists at Microsoft, for instance, debated the internet business question so long that Google left them in the dust.

"But how can I be sure? Isn't a comprehensive evaluation the be-all and end-all?" I hear that from managers a lot. No, not when you analyze it to death:

> You can gather the pros and cons for years—but after a certain point, the basis of your decision is no longer becoming more complete, it's becoming too complex.

Of course, we need a thorough (but not exhausting) analysis. That builds confidence. But if you still aren't sure, you don't need more of the same, you need something different: feedback.

> When you've reached a certain point in your decision making process, feedback is going to give you a greater sense of security than the most comprehensive analysis.

When I'm faced with a difficult decision, at some point I get up, take off my analysis cap and go out into the world to get some feedback:

from my wife, my colleagues, good customers, experts, laymen, God, the world and the guy at the corner newsstand. Even though these people can't really make a judgment call at all—they often see aspects that I might have overlooked or assessed incorrectly. The main thing is: I can listen to myself talk. By trying to convince others of my opinion, I recognize more easily which of my arguments are a bit weak, where I'm in danger of giving in to delusions of grandeur, which aspects I was underestimating and where I haven't thought through each of the consequences.

Does this tip sound reasonable? Then why do so many managers ignore it? They're playing God. But that doesn't help.

Playing God Doesn't Help!

People need other people in order to make good decisions.

I don't know a single manager who would contradict this. Yet many like to play God. They refuse to ask for feedback and make what their colleagues, line managers and employees call "unilateral decisions". Why? Because they think others will see it as a sign of weakness or insecurity, or try to talk them out of their decisions. That's nonsense! For the simple reason that:

You can listen to hundreds of people and they can talk all they want, it's still you, alone, that has to make the decision!

It's really the other way around: when you ask people for their opinion (not their advice!) on an upcoming decision, they won't see this as a sign of weakness, they're much more likely to see it as a sign of self-assurance. Only decisive top managers act with such self-confidence. It's also pretty obvious that they don't approach it like this: "Oh God, I don't know what to do! I'm so confused. Please come to my rescue and make the decision for me!" Instead they say: "We could introduce the new product based on its price or on its exclusivity—what would make you want to buy it?" Then the person you asked feels like you're taking them seriously, they feel involved and say: "Well whadda ya know! The big boss asked me for my opinion today!"

After a few weeks of hard training, most managers have gotten used to questioning those around them, before they make a decision. A few will still say to me: "I don't need feedback. The numbers speak for themselves!" Maybe you've seen through that already. This kind of decision-making approach is also pathological. I'd like to spare you the trouble.

Don't Put Your Faith in Statistics!

Of course you need the numbers for every decision. You need to know: Will it pay off? Are the resources that you need available? Is it technically feasible and do you have the right people for it? The only problem is:

I have never seen financial planning that was negative.

No one is going to come to you and say: "Let's get product X on the market as soon as possible. We're sure to make a loss!" When a decision looms, the so-called "hard facts" are usually (much too) positive and always make it seem like the right thing to do. Or as Mark Twain put it:

"Lies, damn lies, and statistics."

Statistics, on their own, will often lead to bad decision-making. What will save you from this mistake? Again, it's the feedback principle:

Talk with other, but not just with the people in your own orbit.

Your immediate colleagues and employees are not going to be completely objective, which means: They will tend to support your planned calculations—regardless of how far from reality they are. That's why I don't like to talk with the internal folks who can't see the forest for the trees. I go outside our company to talk with customers, industry experts, and sometimes even competitors or critics. If the numbers really do show some weak points or are misleading in any way, then I'll get to the bottom of it much more quickly this way.

This feedback doesn't necessarily mean I've got to toss my plans out the window. But I will definitely feel more confident about where further questioning is necessary and what additional risk management measures should be taken.

Don't be a slave to statistics!

If all the facts are telling you to go for it and your gut feeling, alone, is starting to grumble, then as a veteran manager you're allowed to, in fact you should, say: "Thanks for the exhaustive data collection. I'm making this decision to go against it in spite of the promising factual landscape, because I'm convinced that this isn't going to work." Most employees and colleagues will agree. If some express scorn, you can always say: "I'll sign this immediately, if you present me with some reasonable risk management measures!"

Remember: Even an absolute "no" counts as a decision and you don't have to apologize for that. Decisions require strength. Be strong enough to just say "no", every now and again. It might feel difficult at first, but afterwards you'll feel a lot better than if you'd given in too quickly and worried about the negative consequences for weeks afterwards.

Collective Sin

I was responsible for a large IT project at a company I was working for several years ago. We were planning the migration from a national to an international IT system. I felt very confident about my decisions: our management executives had called in a few of the most renowned external IT consultants to our team.

They analyzed the situation, demonstrated several possible solutions, gave us hundreds of recommendations based on years of experience, but when the time came to decide on one of four possible system variations, none of the highly paid and renowned specialists were able to say it should be "A", "B", "C" or "D". We could have coaxed, cajoled or provoked them all we wanted—we didn't get a peep out of a single one of them. It was like being in Dante's Inferno: I wanted to set something in motion, but wasn't able to because these "house of cards" technical architects didn't want to stick their necks out!

I know countless managers who have had to bake in this kind of hell: they have their kick offs, team meetings, get-togethers and briefings—but as long as they are beating around the bush, nothing comes to fruition. Why is that?

It's a management sin to rely solely on collective agreement. Decisions can only be made by one person!

This is not the same as a dictatorship: the solitary decision-maker brings in the opinions and expertise of everyone involved any way, right at the beginning. That's the way I did it, too. But when the collaborators couldn't come to a decision, I banged on the table and said "We're going with 'C'!" Were they offended by my seemingly emperor-like demeanor? On the contrary: The huge sighs of relief nearly blew the papers from the table. Every one of them was thinking, "Finally the decision-maker has made a decision! That's what we're paying him for at the end of the day! How long was Mr. Schuster planning to wait?" I admit, I felt as lonely as a polar bear in the Gobi desert right then. But:

Making decisions is lonely! A good manager can take it.

Let's be honest: one of the reasons that committees get set up so quickly is that none of the individuals involved want to take the blame afterwards when a decision turns out to be wrong. For people earning a manager's salary that is pretty wimpy behavior.

Enjoy the Solitude!

I know a very decisive top manager who not only deals well with solitude, she loves it. Managers have a tendency to look for consensus. After a bad decision, they want to be able to say: "Come on, you were all behind this, as well!" This manager doesn't need that. Sometimes she agrees with the opinions of her committees. And about the same amount of time she makes a solitary decision, in which she says to them: "You are all against this recommendation, and I thank you for your honest opinion. We're talking about a strategic executive decision, however, and I'm the one who decides. And I am deciding for this." Against the majority.

Usually, no one mutinies. Because everyone is glad that someone had the nerve to put their signature to the declaration. And her decisions usually pay off. Her success is the proof. The courageous director is not a gambler, she only takes calculated risks. That is bravery.

It's the home of the brave they sing about in the American national anthem, not the clever.

The Home of the Brave

I'm sometimes appalled at the amount of fear in management. When I coach timid managers, it usually turns out that most of them don't even know what they're afraid of. OK, making the wrong decision. But we're not talking about Typhus or the Plague, here. Your people aren't going to die if you make the wrong decision. What are you so afraid of? Ending your career, losing your job?

If you think: wrong decision = loss of job, here's the silver lining: this equation is faulty.

I don't know a single manager, regardless of his/her place in the hierarchy, who has been fired because of a single bad decision. If that happens, then it wasn't the first and only mistake, it was the last in a long line of blunders. Furthermore, every manager is going to land in the dog house at least once! Afterwards, time goes by, the dust settles, a few good decisions are made—and the manager can afford another misstep. Ergo:

A little less fear and a bit more courage are good for you and your decisions.

Avoiding risks is a much greater risk. Each of your bosses has also gotten it wrong at some point, so that's forgiveable. Sitting around and avoiding making a decision isn't. That won't lead to success.

Trust is a Risky Business

What's the biggest risk factor in making decisions? Your own employees! They make the necessary preparations and gather the relevant information. I still remember how I good-naturedly okayed all the little investment decisions of a subordinate. We had agreed that there were some necessary investments to be made, in spite of the need to economize. And then I found out that his idea of "necessity" and my idea of "necessity" were poles apart:

Trust is a good thing. But when it comes to decisions, you need to know whose judgment you can really trust.

Make sure your employees are using the same criteria as you when making a judgment call. Only then can you trust in them, completely.

Making Decisions under Pressure

I don't need to tell you that the worst decision-making sins are made under time, cost and performance pressures. The clock is ticking away, beads of sweat are forming on your brow and you just don't have the time to say "Come on in and sit right down, let's take a look at the numbers one last time."

There is only one thing that helps when decisions need to be made within seconds: listen to your gut instinct!

Modern brain research confirms this. Instinctive decisions beat intellectual ones hands down—especially when they have to be made within a few seconds. Malcolm Gladwell devoted an entire book to this subject in *Blink!*—and landed on the world's bestseller list. If we are under extreme time pressure, our intuition and gut reaction is unbeatable. If you have more time, the opposite is true:

If you only have a few minutes to make a decision, there are two things that help: quickly ask those around you—and then make a decision based on your gut feeling.

A helpful criterion which is seldom used, in spite of its usefulness, is this simple question:

Would you bet your own money on this decision?

As soon as the time pressure is reduced and you've got a few hours to think things through, your gut feelings can go haywire. So once you've got time, you should go through the process described above: a well-founded, compact analysis, quick round of feedback, brief optimization of risks, then decision. If the situation is still not clear, go back and ask your gut.

First Things First

Imagine you have to make ten decisions today: Which one would you begin with? With the most pleasant or easiest? Thanks for your honesty. I would too! Perhaps not always but way too often. We tend to put off the most unpleasant decisions the longest.

Making the decisions in the wrong order is another big sin that managers like to commit.

The keyword here is: prioritize! Go through all the upcoming decisions, prioritize them—then tick them off from one to ten? Turn the old procedure upside down and start by making all the most unpleasant decisions first! Yes, that takes a lot of energy and courage. But didn't we just agree that it's the home of the brave?

Just Do It!

Do endless discussions drive you crazy as well? A bank manager hit the nail on the head when she said, "I would rather make a blatant mistake than sit around listening to this nonsense any longer!"

You don't have to wait until "we're all in agreement" before making a decision. You're never going to get to this point in complex decisions—that's why they're called "complex".

Say instead: "In the next five minutes we will either agree to do it—and then do it. Or we will decide not to do it—and never mention it again!" It's a lot easier to deal with a short, sharp shock, than a pain that never goes away. The opportunity costs of endless discussions are high: If you discuss endlessly, you end up hitting a brick wall, only managing to prove that you did, in fact, try everything that you possibly could.

What's the Best Strategy for Making Decisions?

An international comparison helps provide a balanced answer to this:

- The Germans, for example, define a goal. They analyze, plan endlessly and stick slavishly to the plan, regardless of how much the world has changed in the meantime.

- Americans, on the other hand, define a goal, get started without a lot of planning, get into trouble, and keep redefining the goal along the way until it "fits".

Which way is better? Who has the better decision-making strategy? Neither. At the end of the day, it takes the same amount of time to reach the goal. While the studious Germans are still planning intensively and also losing a lot of time, the valiant Americans are starting to make achievements—but then run into trouble, and lose the same amount of time. You won't find the answer to the question of which way is better in practice or in academics. The crux of the matter is:

> Become aware of your own decision-making strategy so you can figure out its pros and cons and then find a balance. It's not about making decisions based on formula XYZ, because "we've always done it that way."

I hear a lot of German top decision-makers saying: "Let's cut down the planning phase as much as possible and just get started—before the others get ahead. We'll go with a parallel strategy: plan and do." While a lot of astute Americans are also saying: "Stop! Don't take off without first planning a bit. Otherwise the Japanese and the Germans will be laughing their heads off when we land head first in the mud."

Best Practice

The best tip that I can give you for making decisions is the following:

To make 'good' decisions, every manager needs a sparring partner.

The ex-chairman of a German court, for example, always discussed his decisions beforehand with his wife, who was also a very experienced and practicing judge. He valued her judgment but much more than that, he appreciated the chance to speak freely and openly in an atmosphere of understanding where he could test out all his pro and con arguments. Such an atmosphere helps in good decision making immensely—even if, in my humble opinion, management gurus and academics have missed the boat in this regard. It's important to listen to the opinion of an absolutely trustworthy person, who has the good sense to say: "enough is enough, you're going a bit too far overboard, now." Or, "Stop beating around the bush, you know the right answer, already! Who are you still trying to convince?"

I can say from my own experience: everyone needs a sparring partner. Most will find one right under their nose. Get one, be good to them, invite them over for a beer or a glass of wine now and again. Sometimes a professional coach is your best choice. It's just like in boxing: you get better with training.

The Chapter at a Glance: Take the Shower Test!

When you jump into the shower in the morning, ask yourself: "What decisions will I make today? Have I got all the data I need? Have I calculated the risks? Do I need to pluck up my courage? How can I do that?"

Demotivate Your Employees!

"I don't know if it's possible to motivate employees. I'd be happy if my top executives would stop demotivating them."

Owner of a mid-sized company

My Boss is Getting on My Nerves

When did your boss last get on your nerves? Oh, I know, we all get along swimmingly with our bosses, officially! But, really: when did he (or she) last get on your nerves?

It's pretty sad, isn't it? Every management job description states that they are expected to motivate their employees. So why don't they do it? Why do they always have to be such a pain in the neck? Don't they like us? Do they do it on purpose? I can give the all-clear on that point:

> No manager demotivates their employees on purpose. It's much worse than that: they usually do it without realizing.

My bosses have also annoyed me at one time or another. And I don't mean this as a criticism. I don't harbor any bad feelings. And yet: it was really annoying that some were not able to motivate me, though it was actually their job to do this. Instead they made careless and unfair remarks, jumped to the wrong conclusions, made decisions without warning, were inconsistent, lacked know-how and/or transparency, displayed high-handedness or just failed to communicate. The usual demotivators.

I found it hard to believe that my bosses didn't notice each time they were effectively giving me a good punch to the gut. I assumed they must know how mad I could be with them, and would, therefore, be working the next few days with the handbrake on. Early on it was clear to me, I didn't want to manage like that! I didn't want my employees talking about me behind my back, calling me "buzz killer" or "arrogant ***-hole" (favorite nickname by demotivated employees for demotivating bosses). I wanted them to respect me. But I won't avoid those nicknames if I am being—inadvertently, unintentionally, or unconsciously—demotivating. And yet I still manage to do it sometimes, even today. As do all good managers. Don't believe me?

How I Got On Someone's Nerves…

My youngest daughter Tina studies business and works part-time at our company. Sometimes she comes into my office asking for some help. As her line manager, I also want to be a good personnel developer, so I pressured her, recently, to look for her own solution and work out two or three options, before coming to me for the answer. At the end of the day, I just wanted my girl to learn to think for herself. If she wanted to become an executive manager, she was going to need this capability. And, do you think she thanked me for my fatherly guidance?

Not one bit. Instead she complained, "You can be really annoying, sometimes. If I already knew the answer, I wouldn't be asking you. Would it kill you to stop with the guessing games for once and just tell me the answer?" Does that sound like gratitude? No, it's feedback, which is more valuable than a pallet of gold bars. It's just too bad that not all employees get the chance to give such honest feedback when their bosses demotivate them. How did I manage to annoy my daughter so much? I meant well and my way of helping was absolutely justifiable and even necessary. Maybe you've already guessed how:

If you focus on the task too much, you're already being demotivating.

That's hard to hear, of course. Because it's our job to assign tasks, isn't it? Unfortunately, we see over and over again how well that functions in reality: and that is… not at all! Although it's our employee's task to take care of things, if we only focus on the things and forget the people, they'll start to rebel.

Ask yourself regularly: am I too task-oriented? Am I demotivating my employees in this way?

Or am I trying to be like the bumper-sticker: "have you hugged your employees today?" I only had one thing on my mind when I expected my daughter Tina to make her own suggestions: "That girl has to learn to think more for herself, otherwise she'll go under in the business world faster than the Titanic!" That was the task. How Tina might react to that hadn't occurred to me for a second. That's exactly what demotivation is. *Motivation*, on the other hand, means offering the task in such a way that the other person wants to do it.

"You shouldn't slap the truth on like a wet cloth to the face, it should be held out like a coat that you can easily slip on."

Friedrich Dürrenmatt

For example, like this: "I know the answer to your question, Tina. But if I tell it to you, it'll be like at school—in one ear, out the other. I think it would be more effective if you would work out your own proposals. Because the most important thing in business is being able to solve problems on your own."

The naked truth is demotivating.

Many managers don't know this. Others don't want to believe it: "But the facts speak for themselves!" say engineers, scientists and information managers indignantly. Yes, they speak for themselves, but unfortunately not for the employees! That's what is meant by managing. If it was all about assigning tasks, we would only need engineers and not managers (nothing against engineers). Are you getting the creeping suspicion that your management style is too task-oriented? Does the following statement apply to you?

"I most like having a boss who treats me like a task to be accomplished."

Even bosses, who manage with "facts, facts and more facts!" in their heads (a very dehumanizing statement, by the way) would hardly agree to that statement. They don't want their bosses treating them the way they treat their employees. Gaining this perspective, as painful as it is, is already half the battle.

The other half is not to just rely on the realization but to actually change your management style. For most intellectuals, nuclear fission is easier. That's why we need a little warm-up:

A Little Brain Jogging

Please reformulate—in your head or with a pencil—the following "task-oriented" statements, so that they address both the task and the person:

1. "Bob, please finish order number 298, today."

____

2. "Kathy, you didn't achieve your monthly goal again."

____

3. "Why are the calculations taking so long?"

Why are these three statements so (unintentionally) demotivating? The first is a fact, it's even polite! However, pure instructions aren't motivating—that's also true for the army, where they supposedly only get commands. If your employee feels like he's nothing more than a robot subject to your will and command, he'll react with automaton obedience, not with motivation. Statement number two is a clear accusation, without any offer of support. Anyone receiving this kind of comment might think they were being left out in the cold. Of course, you can choose to give feedback in this way, but you shouldn't expect your employee to be motivated. And statement three sounds like a three year old in the back seat of a car: "Mommy/Daddy, are we there yet?" As every parent knows, that's also not motivating.

How would you rephrase the three statements? There is, by the way, no right answer. There are dozens. The right ones are the ones that consider the perceived or assumed sensitivity of the person on the receiving end, for example:

1. "Bob, I know that you've got a lot on your plate at the moment, but I need order 298 by 5 o'clock today. I'd appreciate it if you'd do me this favor."

2. "Kathy, you haven't achieved your monthly goal for the third month in a row. What type of coaching or support would help you reach it next month?"

3. "I need the calculations as quickly as humanly possible. Is there anything you could do for me? Really? Could it even be a bit quicker than that? Thanks, that's helps me a lot."

If you can identify the principle behind these model answers as "applied politeness" or simply "empathy", I wouldn't contradict you. A coachee and managing director from the IT industry expressed it (remarkably) as follows:

"I've got the following program loop running through my head at all times which asks: What words do I use to phrase the task, and what words do I use to speak to the person?"

The task and the person—if you address both, you motivate.

Loud Employees are Good Employees

I'm happy to report that I haven't completely failed as a father: Tina is still confident enough to give me honest feedback. That would tend to speak in favor of my motivational abilities because only motivated employees will open their mouths. And conversely:

Nothing motivates employees as much as being allowed to open their mouths …

Freedom of speech in the presence of the boss is super motivating. Of course, it's painful when an employee (such as my daughter) gives us a verbal kick in the shins. And the truer it is, the more it hurts.

Motivated employees are sometimes a pain. A good manager can take it.

Now we know why some managers cannot and don't want to motivate: they can't handle motivated employees. Good managers, on the other hand, think: "no pain, no gain". Of course, it's harder to get honest feedback than to give incentives for blind obedience. Freedom of speech is ten times more effective—it's not in the constitutions of most modern countries for nothing. Furthermore, about 90 percent of the time, motivated employees are a lot more fun—and they'll give your external, and internal, competitors a lot more grief, too.

Most employees also recognize that incentives and bonuses are just a way of making up for weak motivational skills on the part of management. The production manager of a plastic processing company phrased it as follows: "What use is a trip to Hawaii as an incentive, if the boss never listens to me? The trip is over in a week, but the boss will keep ignoring me for the next fifty weeks!"

Ask your employees for their open and honest opinions (you can always work on their communication skills at the same time). Reward them for that, even though it sometimes is painful. That is a more powerful motivational tool in the long run than any incentive.

You will also find out a lot more. Things that you would normally never find out about if your employees are keeping mum out of sheer demotivation. Why do managers have such a hard time granting the liberty of free speech? They are suffering from a phenomenon I call the Beethoven syndrome.

Managers Don't Listen

A divisional manager from the electronics industry once explained to me: "I try to keep my divisional director informed about our big project. That's more difficult than it sounds. Recently, I managed to get in touch with him by phone at an airport after three days and twelve attempts. He let me talk it all through and I said goodbye in a good mood, happy that he was interested in the project. Then he answered with: 'And thanks for calling back!' It became clear in that instant that he hadn't been listening to a word I'd said. He was obviously totally lost in his own thoughts!"

The divisional manager never called this director again. The project is not going well, now. "It's not so bad", she says, "Nobody's interested anyway!" That sounds like demotivation. The biggest value destroyer under the sun. But a company director can't be expected to listen for hours on end, you say?

You can tell your employees: "I'm afraid I don't have a lot of time. But the next five minutes are yours. Can you give me the gist in this short time?"

I heard this example from a group leader in the pharmaceuticals industry: "My boss never has enough time. We call her the 'minute manager.' That's about all she has to give: if we're lucky, two and even three minutes. But in these three minutes she listens to you with an intensity that's incredible. You feel like the center of her universe in that brief moment. She looks into your eyes the entire time, asking questions, making keen observations, wanting to know every detail. She squeezes you like a lemon, getting every last drop out of you. And, if you want to know the truth, none of us can take it for more than three minutes. It's just too intense." This type of "aerobic listening" was perfected by Clinton (Bill, not Hillary). It got *him* to the White House ...

Giving three minutes of exhaustive attention to what you're employees are saying is much more motivating than any bonus you could pay.

The best part is: any manager can learn this! Some learn it in seconds, others in weeks. Some can learn it on their own, others will need the support of a coach.

If you want to motivate your employees right down to the tips of their toes: listen to them with your complete attention—not always (no one has that much time), but regularly.

"But I can't let my employees dictate to me!" is the usual objection. No one's asking you to!

Listening is motivation enough—even if you don't do what they say.

Why is it so hard for managers to listen? Because they are people of action. That's what they're paid for—telling you what to do, giving instructions, being "right", and for organizing.

A good basketball player can switch from offense to defense in a split-second. Try to switch over from talking to listening—and back again.

It's a lot easier for managers who don't think like this: "The average employee—what do they know?" That's an attitude that will demotivate on its own. An employee can sense that from miles away—and the motivation goes right out the door with it. Good motivators know: "Everyone has good ideas: you just have to listen carefully."

If you don't give your people the gift of your attention, you'll end up paying a lot in bonuses and incentives. This won't compensate for failing to motivate. On the other hand, the manager of a support team once told me: "I don't need incentives. I listen to my employees."

One Size Fits All Motivation

What question do people ask most in motivational seminars? You guessed it! "How do I motivate my employees?" I know a number of trainers, driven crazy by this question because it is just plain wrong. You can only answer with a one-size-fits-all recipe, and that's not good: "Treat your employees so and so—and they will feel motivated." Such a recipe doesn't exist. Every employee is different as you have surely noticed already. A general code of practice for motivating your employees is about as believable as the perfect pick-up line.

The right question is: "What motivates employees?"

The answer is simple: It depends! For some it's money, for others it could be attention, or punishment, clear instructions, independence, lots of customer contact ...

Every employee has their own "motivational button". Find it and push it!

That's not manipulation. If you put on the pressure, cajole, threat, bait with cash or a carrot, then you're employing dishonest means of persuasion because you are insulting the true motives of your employees. How can you know what works for an employee and what the right buttons are to push? Look them in the face.

You can judge by their facial reaction whether you've awakened their enthusiasm!

And that is very reliable. Because if you've been paying careful attention up until now, you'll know: "If the employee looks dismayed, it's all over! I've frustrated them!" That's right. And for this reason:

Motivating means mind reading (or at least the thoughts of the person across from you).

And you don't need a crystal ball to do this. You probably know your team better than you think. If one person is obsessed with details, you don't give them a job and say: "Here it is, have fun with it!" You'll see the smoke coming out of their ears in no time; such instructions are way too general. Bombard them with details. From that employee's point of view it's great, and it's totally motivating. If, on the other hand, you treat a free spirit like this, it's more demotivating than spilling coffee on their new laptop. They hate details, they cramp their style. Just give him or her the "big picture" and let them figure out the rest on their own. As the Romans liked to say:

Suum cuique! Each to his own. That's the best recipe for employee motivation.

If you appeal to the individual motives of a person, you will encourage them the most. You also have to understand their current situation.

Made-to-Measure Motivation

A few months ago I was asked to facilitate for a project team which was failing miserably. For hours I discussed options with them for possible solutions. The atmosphere was strained and the team looked demoralized. A young divisional manager who came in to pay us an unexpected visit, noticed this as well.

The misery in the room was palpable, so he decided to loosen things up by taking advantage of popular interest in the latest technical gadgets. He decided not to talk about the project and its problems but tried instead to distract his employees from their worries. He put his own technical wonder on the table and demonstrated everything it could do. How well do you think the divisional manager's motivational efforts went over?

First of all, let me say this: a divisional manager comes to a demoralized project team meeting in a surprise visit. Full points on a motivational scale. Furthermore, he didn't talk about the task, he tried to improve everyone's mood. My hat is off to him! But you probably want to know the rest of the story, so without further ado:

When the top manager had left, one of the leading team members stood up and said: "Why on earth does that clown in a necktie need an iPad paid for by the company? Maybe he needs to watch the football game while we're sitting here until dark on a Sunday, without getting a cent of overtime, to save a project that, thanks to budget cuts made by top management, is now going under. I definitely don't need that." And with that he left the meeting. Another followed him. But they didn't get far.

Through the door came a grey haired man. He was a director, responsible for production, and a kind of legend in the company. He simply asked: "Hey, what's going on, guys?" They began pouring out their frustrations and explained their indignation. He just listened, sometimes asked a few questions, but didn't add a single comment.

After ten minutes, the anger began to pass and the team started developing, for the first time in weeks, some realistic solutions to the problem. How did the old guy do it?

If you want to motivate someone, simply be present in the situation that they are at the moment!

The divisional manager meant well. What he overlooked was:

Consoling, distracting, whitewashing, cheering up—that doesn't really motivate on its own. It doesn't motivate you, either, does it?

The divisional manager wanted to distract them from their precarious situation with the help of his new toy. That's something a lot of people do automatically but it doesn't really achieve anything. Unintentionally, and at a subliminal level, he was really communicating: "Your situation doesn't matter that much to me. At least, it's not as terrible as you're all making it out to be." That puts the brakes on anyone. At another time, the team members would probably have shown an interest in his tablet pc with all its great features, but with the current backdrop of a reduced project budget and their justifiable worries, an expensive toy paid for at the company's expense didn't really go down that well.

You demotivate people, when you don't take their situation seriously.

The director of production did the only right thing: He not only took the people and their motives seriously, he also took their situation seriously. He showed his interest by letting them do the talking. Anyone can do this simple exercise!

I know a board member at a private bank in the south of Germany. Every morning she goes to each of her employees and greets them with a handshake, listens to what's going on in their life, whether it's the baby's stuffy nose, the teenager's tough exam, the new furniture for the living room or the next planned vacation. She never stays that long but the few minutes are enough to get a feeling of what's going on in their lives. Her employees would walk barefoot over live coals for her. You would never do this for a bureaucrat, but you would for someone who took you seriously. You, your motives and your individual situation. When I recommend such an approach, my business-oriented managers often ask me: "What do I say if one of them tells me about something really serious, like their mother's upcoming heart operation?" The answer is simple, nothing! Just listen.

People don't expect advice from you. Advice from the boss can seem like bossiness and doesn't motivate. Listening and letting someone else talk, does.

Show your understanding: "Yes. I understand." "That's a terrible situation." "I would feel the same." "I know exactly what you mean." "Oh man, what can you do?" What answers should you definitely avoid if your employees pour out their heart to you? Right: "Keep your head up! You'll be fine! Don't worry!" These set phrases are about as motivational as a slap in the face. Such empathy-empty managerial answers don't motivate you, if you're really worried about something.

Best Practice

Usually, consultants are called in to throw people out. Cynics call that the McKinsey effect. This enables the board to say: "Sorry, people, we would love to be able to keep you, but the consultants say we have too many so we have to let you go!"

Recently, such a board member actually did call me (and it wasn't the first time): "We have thirty employees too many in administration. Please come over when you get a chance!" I hate to be used as a company wrecking ball (many McKinseyians too, by the way). But because I know how many company leaders think, I said as a test: "Thirty, from several hundred? That should be possible, they'll just have to go, then." "No", said the board member. "I didn't say anything about firing. I want them all relocated to sales positions." I swallowed hard. Hmm, I thought, people who have been stamping forms for the better part of a decade are now supposed to go into the second hardest job in the world (the hardest is being a parent). I asked him what had given him that idea.

"If I let go of thirty people, I will demotivate the rest of the crew, let alone the thirty people fired. And that frustrates me too. I don't want to throw any people out!" He had my respect. Here was a top manager taking "motivation" as seriously as it should be taken. No power in the world could have made him utter the words "they've got to be fired" although it would have been completely ok in business terms. He didn't want to fire them because:

If you only think about business, you're hurting the business!

Admittedly, the dismissals would have been good for his bottom line—in the short term, that is. In the mid-term it would have led to demotivation and could have had a detrimental effect on productivity and the bottom line, as well.

No one can prevent you from thinking about the business as a priority. At the end of the day that fosters success. But if you don't neglect your motivational responsibility, that results in ten times the success.

Word gets around fast when a board member doesn't fire his employees, even though it seemed necessary. That lights a motivational fire that will spread like no other. When I visited that company again, none of the thirty people had become top sales people, in spite of the intensive training—only the Almighty can work miracles like that. On the other hand, everyone is happy with the solution and the people in the sales department managed to incorporate the new employees so that they were able to increase their success quota. They are also just as motivated: "Finally, the board has heard our call for reinforcements!"

Driving the business forward and motivating your employees—keep your eyes on both, and keep both your eyes open!

No one can force you to be an encouragement for your employees. Many managers can be real buzz killers and motivational wet blankets—but they still manage to scrape by because their bosses are also weak motivators. At the same time, they complain that they aren't getting the respect and acceptance they need.

Motivators get respect!

Managers manage. Leaders motivate. Who do you want to motivate next?

The Chapter at a Glance: Take the Toast and Jam Test!

When you're spreading jam on your toast in the morning, ask yourself: Do my employees (or my partner, my children, my customers ...) think I'm motivating enough? How can I train myself to become more aware of how what I say and do comes across? Whose 'motivational buttons' am I going to try to push today? How can I get to know them better? How can I keep up with what's going on in their lives?

Don't Have Any Fun!

"Work should be more fun than having fun."

Tom Peters

Poor, Little Manager

Check the room for bugs and hidden cameras. Are you sure no one's listening? You should also send out the young, the old and the frail, or at least cover their eyes and ears, because we're going to look at a something totally shocking. Let's keep it down to a whisper—the question we're about to ask dares not be uttered in polite society: ... Are you having fun at work?

Accusing a manager of something like that is almost sacrilegious. Because we all know: the world is round, God is big and managers *don't* have fun! Or have you ever seen a top executive really laughing at work? Or genuinely smiling? The best you can expect is usually a forced display of the pearly whites. No, they often look like they want to eat someone's children for breakfast. Self-righteous. Imposing. Deadly serious. Most people think it's the heavy burden of responsibility that they carry around with them that makes them look that way. The truth is: They just aren't having any fun at work.

In spite of their fabulous salary, company car, private jet and sense of privilege. In reality, I only know distressingly few top managers who actually seem to enjoy what they do—in terms of: more fun than adversity. That's weird. You can imagine the reaction of the average Joe on the street: "What? With that salary? The person isn't having fun at work? Man, I earn about a tenth of that!"

Sometimes, when I'm in a good mood, I stand outside the company where I've got an early morning business appointment and watch the parade of "gladiators" marching into work. The workers tend to go through the door laughing, gossiping or dissecting the game from the night before. I can always spot the managers from this crowd. Walking alone, sticking out of the happy crowd like a tarantula in vanilla pudding. You can recognize them by their briefcases, expensive shoes and frowning demeanor. They don't really talk with anyone. They keep their gaze down or look ahead with imperious determination. That seems serious, important, professional, irreplaceable and unapproachable. Like emperors and empresses on parade, stoically carrying the burden of rule and looking like they've just bitten into a sour lemon. But why shouldn't it be this way?

Why shouldn't managers like their work? Surely, you've heard that fun is an important success factor. And you know that no one motivates their employees if they aren't in a good mood themselves. On the other hand: A manager who has a lot of fun at work doesn't live up to the standard that we often want to conform with. A manager having fun? Nutcase. Undertakers and managers shouldn't laugh when they work.

This forced and self-righteous seriousness is a self-imposed penance for that super salary and all the perks that go along with it. Most have come to accept this. The question is just:

Do They Enjoy It?

Don't try to tell me that that's true. Managers are successful. They have a lot of responsibility. Position, status, prestige and salary give them a certain contentment, probably even a kind of grim satisfaction. But fun? Personal fulfilment? That inner glow that they once had when they were young? Few feel like that. So how do you deal with it?

Most think: I have to climb the next rung of the career ladder, get the next promotion, then I can finally have the freedom to do what really fulfills me, what I enjoy, and what makes a difference in my personal life. That's why managers are so competitive—and not, as sensationalistic journalists like to report, because they are career-obsessed, morally infirm bloodsuckers. With every promotion we hope to achieve the freedom that we need to finally do what we've been dreaming of. And so we climb from rung to rung to find out that, surprise, nothing has changed at all! The new position doesn't offer the hoped for independence. Freedom in management is a myth: Managers will always have somebody above them. Even those who make it "to the top" have the supervisory board, shareholders and owners at their backs. They are just as imprisoned in the system as the rest.

Are you managing to have fun? Anthony can't. He's eighty-four years old and the new CEO of his company. Finally, he made it—and, full of doubt, he came to me for coaching. "Are you crazy?" I asked him jokingly, "How can someone in your position not be happy?" He wasn't in the mood for jokes: "I know the place inside and out. Every customer, all the products. Yeah, we could optimize all of it, and expand and that's also what we're doing. But fun? Not really. Not for me. I've just been doing it for too long."

So I asked him: "What would be fun?" His eyes lighted up immediately: "Expand to the East! We have excellent connections there. And there is so much going on in the East right now! Overnight, huge markets are developing." "Well? Why don't you go there?" "Ah, it's the same old story: the daily work, my busy routine, I'm already working sixty hours a week!" OK, that's true for us all. But: Would Richard Branson have said that? Or Bill Gates? Or any other manager that you know that has tons of fun at work and is super successful? You can think what you want about big shots of this caliber—one thing is for sure: they are having the time of their lives! Not in spite of the work, but because of it. Why?

Have the courage to make your hobby your work! That makes your work more fun.

Both Branson and Gates did explicitly what they wanted and enjoyed. Not always, but as often as possible.

Just Do What You Want!

Anthony's hobby is expansion to the East. Since he now has the courage to own up to this, and no longer uses "the daily work routine" as an excuse to put it off, he has developed one idea after another of how he could approach the project. He still hasn't found the time for a large campaign but he has already started talks with a potential joint-venture partner in the East. It's not a quantum leap for his business (compared with the "right" market entry), but it's a big step for Anthony: "The fun is back!" he says in coaching and grins from ear to ear. His wife has also thanked me: "I don't know what you said to him in your coaching sessions, but my husband has finally stopped moping around the house—that was really damaging family morale. Thanks!"

> You want to have fun? You don't need a higher position for that. What you need is more courage.

Be brave enough to regularly do (also in preventative doses) what you really love. Of course, you shouldn't let your hobby get out of hand. I know board members who almost exclusively worked on their own "board room projects" (meaning personal enjoyment), because their jobs stank so much. That's taking it to the extreme, of course, and is not going to help you take care of business. However, you only need a little time every day to devote to the things that will help you get your groove back. Not everything is going to, or can be, fun but this will help you become a better manager.

> How high is your fun-o-meter at the moment? Everything still A-Okay? Or is it high time you devoted yourself more to your hobby?

At this point the wet blankets are going to ask with alarm: "But is a manager allowed to have fun at work?"

Fun is not a luxury, it's a necessity!

If bosses aren't having fun at work, they're going to lose their employees pretty rapidly. No fun at work means working with the handbrake on—that's a clear violation of the employment contract. It doesn't take long before success goes out the window as well. And every burn-out begins this way.

Is fun in management allowed? It's not just allowed, it's a must!

And by this I don't mean the fun that has nothing to do with the actual work at hand. Unfortunately, top managers are somewhat notorious for this and often get in trouble with ethics committees: Brothel visits paid for by the company, taking bribes, embezzlement, tax evasion. For some managers that's fun but it's actually just a waste of time: It won't make you enjoy your work more or make you more successful. Fun through distractions blocks your performance and is not much fun in the long run. If you're looking for fulfillment at work, you should look for it there. Not just in what you do, but in how you do it.

Have It Your Way!

Managers can't always do what they want. But they can usually influence how they approach it. Anthony, for example, works four to five hours a week on his pet Eastern expansion project. That still isn't enough for him to really feel fulfilled. Unfortunately, he can't find more time for his hobby project at the moment. That's why he now has a new approach to his other tasks to make them more fun—and there's always time for this.

Do you remember how I described my approach to trouble shooting and analysis? The world is expecting me to be studying Excel sheets all night long and continually demand new projections and calculations from the controlling departments. I could definitely do it that way but it wouldn't be any fun. As an ex-banker I do believe in figures and facts but when I have to depend on those things alone, I start getting itchy feet within a couple of hours: I've got to get out to where the people are. I want to experience the business, not in figures, but "live and up close".

So I leave my desk, organize a company car and drive out to the grass roots: to the employees, customers and processes behind the figures. Every time I do that, I'm met with a considerable amount of skepticism—a top manager at the grassroots is seen as an anomaly, that's how far we've come. To me, however, it's the only way to figure out where the fun is and what are the behavioral problems that don't show up in the management reports.

If you want to have fun at your job, and in life, do the things that are the most fun for you. The other things will take care of themselves—and that doesn't have to be what you've been told, or the way that it's always been done before.

Why shouldn't you act like everybody else? Because in management the sheer pressure to conform prevails. The rotten mortgages which brought the world to the verge of a financial collapse were judged to be first class manure by many bank managers right from the start. They bought them up anyway. Because in management, it's better to make a mistake together than be the only one who's right. But if you give in to this pressure, you're going to pay a high personal price:

If conformity is more important to you than joy at work, you will end up very unhappy (and enjoy little success).

That is what a self-determined life is all about: It is only truly filled with happiness when we are self-determining.

If you want fulfillment at work, you should devote yourselves to the things that give you joy. Everything else should be done in such a way that it can also be fun.

"I can have fun after work", said a successful head of department from a pharmaceutical company not long ago. "My career is more important to me than fun and games at work." I would beg to differ.

The "Life of the Party" has the Best Career Chances

A good friend of mine, a successful auditor at an insurance company, hit the glass ceiling at his company within a short time for exactly this reason. He confided in me: "In auditing, no one has very big career chances. I'm switching to something different. In sales, you can make it to the board." It's just a shame that now he would not only be juggling columns of figures, he would also have to motivate his employees. That didn't really make him happy and his sales people noticed it. He still hasn't made it to the top yet.

Having fun at work is still the best career booster.

But fun can do a lot more for you than rev up your career. It can change your life. Let's look at the example of:

A Man Who Could Sew

Since the age of fifteen, my brother has been a salesman with heart and soul. Sometimes he calls me up, audibly pleased: "Do you remember the …? I just sold the washing machine from … with the super savings program!" Think about it: he's excited about a washing machine! I could understand if it were a tablet or some other digital gadget. What's sexy about a washing machine? I don't know, but my brother does. And he loves it. Still. After more than twenty years. That's why he has become the top salesman in the region. If you're having that much fun with something, you can't be anything other than very successful.

That doesn't mean that his (work) life is always a bowl of cherries. Quite the opposite. "Fun guys" have the same challenges to deal with that a sourpuss does, they just deal with it differently: when my brother was sixteen, his boss decided to sell sewing machines in addition to light bulbs, irons, and other household appliances. My brother still tells the story with a voice full of horror: "Unimaginable! I was sixteen, a young man, I had never had a sewing needle in my hand. And now I was supposed to sell sewing machines!"

What did he do? There wasn't a lot he could do. It was a management decision. He had to sell the things, but he hated every single one of them with a passion—and, therefore, managed to sell a grand total of zero in the first few weeks. "I complained", he says today, "to anyone who would listen to me that our machines were too expensive, that they lacked the features that our customers would want, and so on and so forth." Exactly the kind of thing said by salespeople who don't sell a lot. Most sales managers are likely to think that it's the price, the model, or the competition and/or a lack of sales training when the sales targets aren't reached. It would only occur to a few that it's the result of a lack of fun.

December came and all the retailers in the city were setting up stands in the old town Christmas market. My brother's store also had a stand there, with the hated sewing machines. My brother could not believe what he was seeing: His boss sold the things as if they were rubber boats during a flood! What was the sales secret of his boss? It was simple: He demonstrated to the women, as a man, how to sew. Back then, it was a gender coup, par excellence.

My brother was so impressed, he immediately tried to sew something himself, he decided on button holes. Anyone who has never used a sewing machine should ask a sewing member of their family (although seldom, today) how infernally difficult that is. In any case, my brother was a big hit at the Christmas market that year: A sixteen year old boy who could sew so well, he could even manage the high art of button holes! The women, flocking in droves, could not believe their eyes, they elbowed their way to the front with amazement and awe—and bought every last one of the things from my brother. By the way, button holes were the only thing that he could sew …

The moral of the story?

Your Personal Philosophy

Let's assume for a moment, that a lot of things in life are not that fun. That fate is going to throw a few hurdles or sewing machines in our way. We can go ahead and complain about it (which is absolutely human and understandable). Unfortunately, whining has a huge disadvantage: You won't have more fun at your work or in life, as we have come to know through painful experience.

My brother could have taken that approach himself: "If I have to sell the stupid sewing machines, I'll just take more time out for my hobby and devote myself more intensely to selling washing machines and installation material!" And that would have worked, too. Hobbies are always fun. Unfortunately, he couldn't have unloaded the loathsome sewing machines that way.

You can only rid yourself of a 'fun block' with the help of a personal philosophy.

The funny thing is: I don't even know what my brother's philosophy is. One second he hated the sewing machines, in the next he was teaching himself to sew button holes. What got him to that point? Was it a young man's ambition not to let an old man get the better of him? Was it the aspiring hope of winning over an audience? Was it the challenge of learning something no other young man at the time would have dared? I don't know. But, it doesn't matter! Because it doesn't matter what personal work ethic and philosophy another person has. It matters that you have one of your own!

When I tell the sewing machine anecdote and ask coaches and managers for their personal philosophy, they mostly answer spontaneously with, "I would have done it to land a killer deal!" Or, "I would have learned to sew button holes to show my customers what my product can do for them." Or, "It's not that difficult! I'm sure I could do that as well!"

Why would you have done it? That's your personal philosophy! Everyone has one. It is already there, it just wants to be (re)discovered.

> If you stay true to your personal philosophy, you'll have fun. Always and everywhere. You'll be successful, as well—but that's just plain obvious.

So why do we have so little fun in our every day work? Often we let ourselves be overrun by operational urgencies and unpleasant peers, crises and stress. The white noise of our environment causes us to ignore our inner voice which wants to remind us of our personal philosophy. It was like this with Anthony too. He was so busy with his career, promotions and daily business that his personal philosophy got buried. What was his personal philosophy? It was simple: opening new markets, international expansion, acquiring new customers, creating innovative products. Anthony is an innovator, a forward thinker, a visionary. Then he was promoted to General Director and suddenly noticed that he had become a preserver, conserver, administrator, and security guard!

Rediscover Yourself!

What is the difference between Branson, Gates and you? Probably nothing worth mentioning. The great top managers are not geniuses or gurus. They can probably do a few things better than you and I—but other things worse. Those are—*Achtung:* it's the guiding theme of this book - not the outstanding abilities that make the outstanding difference. It is the mistake that Branson and Gates avoid: they don't let themselves be deterred during the day as often from their personal philosophy. They stay true to themselves longer. They live their philosophy more broadly and deeply, more focused and more often. What's that you say? That's not possible? That could spell trouble?

Let me tell you a little story about a cautious and charming middle-aged man who set up a company for secure transportation. The business was going great until 9/11. After that it went crazy. He was drowning in a flood of orders. The boom attracted newcomers to the supply side of the market like maggots to bacon. Soon there was a surplus of supply. Even long-time customers were trying to get him to lower his prices. Desperate competitors had carelessly started a price war. The co-owners of the business were putting pressure on him, his family was begging him, and many employees even threatened him: he should lower his prices to get more valuable orders.

His reply was always the same: "A good security guard and a good truck cost money. If we can't afford that anymore, our transports will be easier to crack than a walnut." He remained tough. He lost some orders. Everyone in the world around him said: "See! We told you so!" He didn't worry. What gave him strength? He stayed true to his personal philosophy: "We guarantee security! All the time and without compromise. That is our highest priority."

While the others ran around trying to get every order possible, no matter how small, and effectively working themselves into the ground with degrading price wars, he kept having fun at work. Because what mattered to him was offering uncompromising security to his customers. As long as he could do that, he enjoyed his work. Other companies gave up in frustration, although they had a better standing than him. They failed, because their owners and managers had exhausted themselves. Because they didn't see the point of fighting anymore. He, on the other hand, kept hanging on. Why? Because he didn't need to "hang on". He was having fun.

When the crooks in the region found out how many cheaply outfitted security transports were on the road, robberies became common place. On some days the roadsides were lined with trucks that had been broken into. Customers ruefully returned in droves—but not at the old rates. Now they had to pay a premium. "That's the price you pay for that kind of stupidity. Hopefully, they've learned their lesson", says the security entrepreneur today with a knowing smile.

If you live your personal philosophy, you will not only survive in times of crisis, you'll also have fun and will be rewarded with success in the end.

Why that is, I can't really say. I have not been able to figure out the connection between fun and success to this day. And it doesn't really matter to me, to be honest, as long as the principle works and I can approach my work successfully.

The great thing about a personal philosophy: You have one even if you aren't aware of it immediately. The less you feel joy and fulfillment at work (and in life) the quicker and more intensively you should start your search—you might need to dig a little deeper. If you're not sure where to start, try thinking about what is the most fun at work and why, that will put you on the right track. The more you let this combination of values and strategy orientation flow into everything that you think, say, do and manage, the more fulfilled you will be in your work, relationships and life. A truly great feeling, I can assure you.

Recently, I met up with an old friend who is a bank director. I asked him, "Hey, how're you doing?" He said, "Oh, OK, I guess." I had to smile, a clear indication of a temporary loss of his personal philosophy. I made an enquiry, "Have you got a shovel with you? We need to do a little digging."

Resisting Temptation

Why are we so often disloyal to our personal philosophy? Because the world is a hotbed of sin to quote Andreas Gryphius (a poet and dramatist). We hear constant whispers that it's all about position, money, power, influence, deals, success, expansion, shareholder value, strategy, change, outsourcing, balance sheets, employability, globalization. It is simple to fall for this trickery and lose sight of our personal nature.

All the Eastern poets and thinkers knew and wrote about the worst sin of all, to be disloyal to yourself.

"This above all: to thine own self be true."

It's always worth following your own north star. I'm not saying this without reason. I was, after all, on the board of an international financial institution. I had "made it". I was at the top of the heap. And the mere thought that I would have to do that until my retirement set off waves of nausea in me, as though a five day old mackerel were being held under my nose: that was the greatest job of my life but it was not the best possible plan for my life.

In my time as an international troubleshooter for my company, I learned what I really wanted, what made me happy and what my philosophy was—I wanted to help companies (and people) achieve seemingly impossible turnarounds, build up new markets, increase my network, see the world, share my experiences with young, striving people, confer with men of state, with poets and thinkers. I was not able to do that to the extent and with the freedom that my personal philosophy longed for as a member of the board. I could only do this as an independent consultant.

Sure, I felt panic before the big jump. Of course, I didn't sleep well in the first few months after my leap into self-employment. Yet, when you follow your north star, you know why you are doing it and that it will be worth it in the end. Because the alternative, renouncing your fate, is terrible. That leads to running into old friends and answering the question of how you are doing with "Oh, OK, I guess." That isn't a nice life, that is not happiness, and it is not fulfillment. That is hell and the reason why two thirds of all managers have alcohol problems or medication addictions, or are sociopathic or neurotic. Or as Henry David Thoreau said, "Most men live lives of quiet desperation."

I probably don't need to tell you that my jump into cold water was worth it. My new team of consultants did brilliant work. I have the freedom to do what I want, when I want, including climbing mountains if I feel like it. I'm having the time of my life. What more could you ask for?

> *"It's a helluva life—if you hit it right."*
>
> Alan Arkin

What you are holding in your hands, right now, is the hard copy of a hedonistic lifestyle. Can you imagine how many publishers were filled with indignation at the thought that I wanted to write a book for managers based on my life and my experiences? "That won't work! Management books shouldn't be written in the first person! Management books have to be based on facts. They have to be dry, serious and academic. After all, it's not all about managers having fun!" It would have been simple to write this book as a third person treatise on objectivity—luckily, there are still publishers who support and understand both facts and fun. Not to mention that the readers of my columns would have been amazed—and bored—if I had stuck to the third person. But even if I could, I wouldn't have wanted to. It wouldn't have been fun.

For you either, for that matter. You see: the choice to find joy and fulfillment at work and in life is a question that you have to answer every day, every minute and every hour of your existence. We can do things the way we have always done them, like the boss or the humorless lecturer wants. Or we can take the path that seems right and fulfilling to us. We have the choice.

The Chapter at a Glance: Take the Shoestring Test!

The next time you're tying your shoelaces, ask yourself: What do I enjoy doing most? Why? How can I get more of this into my life (related to work activities and not escapist)? What's my purpose, my personal philosophy? How can I get this to flow into everything I do, think, say and undertake, as much as possible?

Talk in Riddles!

"Can somebody translate what my boss is saying?"

Joachim H., Department Head

Are We Speaking the Same Language?

Can you understand your boss? Just this question triggers amusement in seminars and coaching sessions. Does your boss come across like an alien? Could he be drunk? Or speaking a foreign language? No? Then why doesn't anyone understand him? Even worse: why doesn't he notice that no one understands?

Some managers believe that just because we are speaking the same language, we will automatically understand each other.

How often misunderstandings occur, should be obvious to anyone who has ever been married (or spent time around a married couple): Couples often talk at cross purposes, although they are speaking the same language. In a relationship, the damage won't usually kill you. In business life, the misunderstandings can be very expensive. Here's an example:

Before going on vacation, the sales boss of an industrial goods company said goodbye (which he didn't always manage to do) before going on vacation. About half way to the Dominican Republic he called one of his busy salespeople and said: "When Schmitz calls: please give him preferential treatment!" The sales boss went away, in came the call from Schmitz.

When the Head of Sales got back from his well-earned vacation, his shouting could be heard by the doorman at the other end of the building. “Are you crazy?!? How could you give the order from Schmitz priority over our other projects?” “But Mr. Withertillermiller”, stuttered the chalk white employee. “You said that I should treat him with preference.” “But I certainly didn’t mean for you to put the brakes on our top projects!” shouted the Head of Sales. “You were just supposed to make sure that his order wasn’t left sitting for two weeks in processing like last time!”

If he meant that, why didn’t he just say that? How can a guy who’s earning about 200,000 dollars per year, express himself so enigmatically? Go figure. The company suffered an enormous loss because a top manager wasn’t able to articulate himself clearly. Because he was talking in riddles. How crazy is that? Isn’t it a hiring prerequisite that all new managers at least master their own language?

Most bosses receive the salary of a boss but they don’t speak the language of one.

A lot of people think of managers as watch dogs barking at their poor employees 20 times a day to show them who’s calling the shots. A majority of managers also think of themselves as being “very direct and clear in my directives.” When I ask their employees, however, whether they understand what their managers have explained to them, the above mentioned amusement pops up again, often coupled with jibes such as “My boss likes to talk, I can tell you …” “Nobody understands our boss, not even his wife …” “I don’t think she really knows what she wants, to be honest …”

When managers are not able to express themselves clearly, why don’t they get some training? For one simple reason: They don’t think it’s their fault! They think it’s the employee’s responsibility to understand: “My employees know what I mean!” intones the sales boss from our example above, with heartfelt conviction. How nice:

The manager can't explain himself, so the 'stupid' employee is to blame.

Does that sound like your boss? Perhaps you also like to make fun of their inability to get their ideas across. Just think about what your own employees are saying about you … It's not amusing, anymore? Don't worry, there's help. And it's not training in elocution. That won't help, here.

Communication is Attitude

To communicate effectively, you don't need training in elocution. You need the right attitude.

The aforementioned sales manager's attitude was, "You all know what I mean! You people should know these things on your own! That's why I hired you!" If you think like this, you're setting off an inevitable chain reaction of misunderstandings. And that's guaranteed. As a manager, I wouldn't admit to such a statement. I would then be saying, "The receiver is responsible for understanding my message. If my employees don't understand what I say, it's their fault." In fact, the opposite is true. If you speak in management riddles, it's your fault!

If you're going around with the wrong attitude, there's only one thing that can help you: Take yourself into the repair shop, change the oil, turn some screws and adjust the settings. A coachee once told me about his own tune-up: "I always assumed that my people knew what I was talking about! I mean, they're smart people. I don't want to insult them by going into more detail!" He paused for a moment, focusing his gaze on me. "In the last two weeks I did an experiment and threw this assumption 'out of the window'. I replaced it with the assumption: They don't understand me at all!"

If you know you're not being understood, you'll be understood better.

If you make the assumption: “My people know what I mean!” then you are doomed to repeat everything hundreds of times, before they finally begin to do it right. If, on the other hand, you assume that your instructions will not be understood, initially, you’re going to communicate, differently. You will check for understanding. My coachee carried on: “For the last two weeks I always asked, ‘What will you do now?’ each time I gave instructions or delegated something. My staff would explain their approach to me and I would ask more questions—usually, I found that they only understood 30 to 60 percent of what I was trying to say.”

If you want to be understood, double check.

I pretended to protest: “But then you’ll have to talk with your people for hours! No one has that much time!” He smiled, “Of course, I have to spend a few more minutes talking with my employees. But I save hours and days of wasted energy: my staff do it right the first time. They’re also a lot more motivated, because they notice that their line manager has some time for them!” He managed to gain this insight by assuming that no one would understand him correctly.

Check your Expectations!

The COO of a large bank instructed his Head of Operations to reorganize the central processes of the bank. A huge project. The Head of Operations replied with a sigh: “Yes, our core processes are totally inefficient.” Then he implemented a new IT system, which would redefine and document all new processes. Project duration: three years. When the cost to income ratio worsened dramatically during the first year (the project required the comprehensive procurement of new IT equipment) the Head of Operations was summoned to the COO. What do you think he said?

Because we are dealing in this chapter with “Misunderstandings in Management”, you’ve probably guessed the answer already: The

Chief Operations Officer had meant something quite different when he said "Reorganization of the Core Processes". The Head of Operations still doesn't know what that was, or me either for that matter. We only know that the COO exploded at that moment: "I didn't want to create new costs! You were supposed to reorganize in such a way that we achieved quick wins in the first year!" The Head of Operations simply asked, "Then why didn't you just say that to begin with?"

The toughest competition in the world won't do as much damage to your company as poor communication from your internal managers. I have to admit, I was also angry at the Head of Operations when I heard about his mistake: "What was that guy thinking when he ruined our P&L like that? Hasn't he ever heard of quick wins, or at least cost neutrality?" To that, my wife said something (as usual) very clever:

"You can't expect others to think the way you'd like them to!"

But that's exactly what we do when we talk in riddles. Therefore:

While (and before) you talk with someone: think about your expectations!

Above all your "have to" assumptions: "He/she has to be able to see that, clearly!" "He/she must be able to understand that!" Normally, the person you are talking to doesn't "have to" understand anything at all. You're halfway there if you've got your own expectations under control. The other half is simple and easy clarity.

Managers, speak clearly, precisely and thoroughly!

"We have to improve our order lead times." Just hearing that drives me crazy! Is it clear? Precise? Thorough? "I want to get 80 percent of all orders for special tools processed within two weeks from the receipt of order to billing." That is clear. However, we still don't know how to do it *ad hoc*. And while we're at it: What will you say (to whom) next? And how? Is it clear, precise and thorough?

Great Expectations

Disjointed expectations are at the core of many misunderstandings.

If you have your expectations under control, you communicate better. When I was called in, a few years ago, to help with the crisis management of a regional bank, my chest swelled with pride overnight: Me! The great crisis manager! I would be flying around in the company jet at all hours of the night and day, fighting fires in the businesses and branches of the company around the world. Alas, it was all just a beautiful dream.

In reality, my colleagues and I spent ten hours a day at a desk, reading papers, playing Tetris and drinking coffee for months, waiting for the big crisis that never came. I must have really gotten on the nerves of my line manager back then (a belated and sincere sorry!). Communicating with me back then was about as pleasant as a root canal operation without the anesthetic. Once I even said to him: "Either give me some work or fire me!" He, on the other hand, tried to explain to me that it was not possible: If he gave me a project and a crisis broke out somewhere, suddenly, I would be committed to this other project and wouldn't be able to leave. We got on each others' nerves like this for quite some time. Until the light finally went on:

If you have unrealistic expectations, you will not communicate effectively.

I had expected, as a big crisis manager, to work twenty-four hours a day, seven days a week, jetting around the world at a moment's notice. Because I had not given this expectation a lot of thought, I got on the nerves of pretty much everyone around me. Much like the line manager that expected his employees to somehow know what he wanted from them (even though he didn't say it).

The quicker you align your expectations with reality, the more effectively you will communicate!

And the longer your marriage or relationship will last (if you have one).

The Broken Manager-Marriage

> "Sorry, honey, I'll be a bit late tonight …"
>
> "Not again!"
>
> "Yeah, the board just called …"
>
> "But we had plans to go to the theater tonight. We've got great seats! And the babysitter just arrived!"
>
> "I know, I'm sorry, I'll make it up to you this time. Promise!"
>
> "If only I had known that you would never be home, I would never have given up my job to bring two children into the world!"

Two years later she asks for a divorce. Why? Because they were not speaking the same language. They talked at cross-purposes instead of discussing their different expectations. At the altar, she expected that the family would come first. He assumed that she would understand that his job had to be priority number 1. In my experience, the divorce rate among managers is significantly above that of the general population. People rarely express their expectations, even in marriages where neither are managers. Do you know what your partner expects from you? Does he or she know what you are expecting? Why don't you talk about it?

My wife and I both work a lot. Every two weeks we compare our calendars, set priorities, and make plans. That sounds pretty unromantic, but it saves us from some disappointing telephone calls. It's not really as heartless as it sounds. What could be more intimate than sharing each other's ideas about how to live together? That creates a bond. It's much deeper than the ordinary pledge of "I love you", and then seconds later the executive board calls, and the spouse and children are forgotten again ... Why do so few couples talk together about their expectations? Many believe that because they know their own desires so well, their partner must be in on the picture.

If you want your communication to be crystal clear: talk about your mutual expectations!

Do you know what your partner wants from you? Do you know what your boss expects from you?

What Does Your Boss Expect from You?

It sounds absurd, but:

Many managers have problems with their bosses because they don't know what's expected of them.

The CEO of a mid-sized company called me up asking for some coaching: "The owner is putting me under pressure. I know that my figures aren't up to standard but I'm doing good work all the same! I travel around about twelve hours a day for the company." These two people obviously had a pretty severe communication problem. We went over her daily routine together.

It turned out that the good man was participating each evening in community projects, fund raisers, and charitable events such as benefit balls. I voiced my surmise that he was also investing a lot of time in these projects during the day. "Of course", he said, "Otherwise it wouldn't work." The coaching sessions generated quick success. It's simple: "Your boss doesn't expect that from you as a priority!"

If you want to understand your boss, you need to understand their expectations.

"But as the CEO I have to be representative of our business as well as show commitment to our social responsibilities in this city!" replied my coachee, a touch on the defensive. "That's right. But that isn't what the owner of the company has in mind. He thinks you should be meeting your performance targets as a priority." First the balance sheet, then a night at the ball. That may sound banal, but when I see what some experienced managers get up to during their work days, I'm certain they've not considered their bosses' expectations.

What does your line manager expect from you? Do you just think it or do you really know it? Did he/she actually say it? What were his/her exact words, and how did he/she mean it? To what extent are you living up to your own expectations—and to what extent, those of your boss?

Corrupt Managers

A lot of people talk about corruption in management. Corruption is only a symptom. The real cause of the illness is quite different, namely, ambivalence towards our own expectations.

Many managers are blind to their expectations.

At conferences I usually stand at the coffee bar where managers with exorbitant salaries show off their latest toys and gadgets. If one of them has something new or great, the others quickly grab for their smart phones and order their assistants to: "Get one for me, too!" At the company's cost, of course. It's just small amounts that the company picks up, maybe 500 dollars or more. That's not the point. The point is: Would the manager have bought the new toy with his or her own money? Why is even this little misappropriation of company money immoral? What is moral, anyway? (You can read more about that in the "Be an Unscrupulous Scoundrel!" chapter.)

It is immoral to violate the ruling expectations. Expectations are norms.

When the board member of a German bank was sued in the Mannesmann lawsuit, he grinned and made a victory sign in front of the courthouse on the first day of the court case. He hadn't cheated anyone or caused any material damages. However, "Victory Joe" as he was called afterwards by his countrymen (with no small amount of sarcasm) went out of favor because he had violated the ruling expectation that a board member should not behave like a teenager. This top manager realized this instantly and changed his behavior, which shows that he really was a top manager:

Top managers have to be aware of the expectations made of them, and always keep them in view.

Of course, I'm aware that the expectations put on managers normally collide with their own imaginations. Owners expect managers to help them make more profit. (Some) managers are speculating, on the other hand, that they will make insane amounts of money, have breakfast with celebrities and enjoy an endless supply of beautiful, well-built talent from their pool of applicants. That's probably true.

It only becomes dangerous when managers lose sight of certain expectations or judge their importance incorrectly.

If you repress your own needs, you'll burn out pretty quickly. If you misjudge the expectations of stakeholders, you'll be out of favor in no time, like the Leesons and Madoffs of this world. It's not about fulfilling all the expectations that everyone has of you. There are definitely some that you can easily ignore. And there are others that you should never ignore. Because if you are not deciding correctly on a daily basis, you'll eventually be caught with your hand in the cookie jar—or with a new smartphone.

What Do Others Expect? Ask Them!

Recently, I met up with a newly appointed divisional manager—and what was the dynamic young newcomer doing when I arrived? He was inputting the names of employees on company access passes. An entry level job! "Have you lost your mind? What are you doing?" I asked. His answer: "My director asked me to do it. She wants to make sure that nothing goes wrong with the new security system."

That was, of course, nonsense. The divisional manager had just not understood the director's management riddles. The director didn't want him to do it himself, she just wanted him to make sure that it was done correctly. Since you can't really count on your boss reading this book himself:

Assume that even directors are not able to articulate their expectations, clearly. Take over the task, yourself, and check the explanations, carefully.

Repeat what your line manager has said again and then double-check it. And don't try to tell me that this is trivial.

This is what a former retail representative in Malaysia thought as well. He asked his chauffeur to drive him to a restaurant in the evening, and then let the good man have the evening free to spend with his family. After the meal, he asked the maître d' to call him a taxi. After half an hour and a couple more shots of whiskey, he asked where the taxi was. The maître d': "At the moment, all the taxis are in use." The retail representative needed to exercise no small amount of self-control not to let out a few choice expletives. As always, the problem was that they hadn't understood each other's expectations. He had expected the maître d' to let him know if there were no taxis currently available, and then inform him of the alternatives. The maître d' assumed that he would wait, patiently, like other Malaysian guests would:

If you have expectations, it's your responsibility to speak up about them.

The retail representative should have said: "Please call me a taxi. If there isn't one available, let me know immediately." Spouses, generally, don't say what they want from each other, either:

> "Why am I always the one to take out the garbage?"
>
> "Well, why don't you say if you want some help!"
>
> "I shouldn't have to say anything. It should be obvious!"

No, it's not. What's obvious to one would probably never occur to the other. This is something that has to be said out loud:

> "Why can't I read the paper in peace, just for once, at the end of a long day!"
>
> "How should I know that you're annoyed when the family comes over to talk with you when you get home from work!"
>
> "But it's obvious that that's annoying! Can't you see that?"

No. And even if it were: One doesn't ***have to*** see. One ***has to*** speak out about it. Before it leads to an argument.

Articulate your expectations every chance you get!

Don't get me wrong, your relationship is your own business. You're free to ignore the expectations of your partner or spouse until the divorce papers arrive. But if a manager is receiving a manager's salary, they should not behave the way that they do at home, they should be articulating their expectations clearly and precisely—and double-checking the instructions that are given. It's hardly any wonder that all coachees report back: "Since I've been double-checking expectations not only at work, but also at home, I've been getting along with the kids much better."

Double-Check and Be Precise

When managers complain that their employees don't do what they expect from them, I let them go on and on with their lamentations and then ask them pointedly: "What? You aren't able to express yourself in such a way that your employees can do what you expect?"

It's not the employee's task to understand the boss. It's the manager's task to be understood.

If you internalize this message, it won't take you long to make it part of your everyday communication. I have gathered together a few recipes from managers, who, according to their employees, do not talk in riddles:

- A financial director confided in me: "I tell every employee when I'm delegating something, 'Please write down what this task involves in your own words.' For even the simplest of tasks there are always details which are not understood. That's normal. What's not normal is not clearing them up."
- I send out an email after every meeting: "Here are the points that we talked about." It's indispensable, both as a memory aid and to ensure clarity. Sometimes I get the others to take the minutes along with me.
- A production manager does it like this: "I often ask my employees to send round an email with the results of our meeting. If I haven't seen any corrections within two days, the minutes are considered correct and binding."
- An office manager, who had admittedly had some coaching, already, reports: "I don't really give instructions anymore. The risk is too great that I'll demotivate them and/or create more misunderstandings. I prefer to use a type of Socratic dialogue. I question my employees on how they would take care of the task

themselves. When their approach is correct, I reaffirm this. If not, I keep asking until they come up with the right answer themselves. That takes a little bit longer. However, misunderstandings are minimized and the employee can do all the work—and more importantly, they're a lot more motivated afterwards because I didn't talk their ears off. They came up with all the possible approaches themselves."

Talk With the Goal of Being Understood!

I know, we all talk rubbish at some point. It takes a while to get into the habit of using understandable, precise language. That is our task in life. Or to quote Goethe: "It takes an entire life to learn to talk like a human being." The following questions help me and my colleagues in the consulting team:

- Who definitely did not understand me today?
- In which conversations could I ascertain a certain insecurity?
- Why? What formulations were not easy to understand? When did I talk in riddles?
- How could I have formulated it more clearly, precisely and concisely?
- Did we only discuss things, or did we agree on concrete measures for implementation?
- When was I able to express myself so clearly that I could see the metaphorical "light" going on?
- How did I manage that? Which general and universal communication rules could I derive from this?

The Word is Mightier than the Sword

"My employees already know what's expected of them." Someone who says that is surely talking in riddles and is doing without a very important resource:

Language is the most powerful management tool.

Recently, I visited a sales organization, which was pretty "trained out" already. They had spent a fortune on advertising and sales support until the sales people were selling in their sleep. Nonetheless, they still closed deals well below the agreed sales targets. I asked for a joint meeting with the sales managers and salespeople. First, I asked the salespeople:

> "Do you ask the deal closing question in every sales talk?"
>
> "Uh, no, sometimes the customer just wants a couple of brochures to compare prices, and stuff."

When they heard this, the sales and marketing managers went nuts:

> "Are you crazy? Why did we just spend a fortune on training to close deals, if you don't even try to close the deal?!"

Before the salespeople could give their usual excuses and explain why some customers shouldn't be asked the deal closing question, I did what I've been asking you to do over and over again in this chapter. I articulated their expectations:

> "Do the sales managers want every sales talk to lead up to a deal being closed?"
>
> "Of course! We're not running a charity here!"

I asked the salespeople:

> "Can you go along with this directive from the sales managers?"
>
> "Sure, of course we can! Why didn't anyone say that before? But how do we ask a deal closing question to customers who just want to have a look around?"

Now, they were talking. Only after the sales managers' expectations had been clearly formulated, could the actual problem be addressed. Namely, that the salespeople had not initiated a deal closing question in (almost) every customer sales talk. The question is just: Why did the managers need an expensive consultant to reformulate their expectations? Why couldn't they do it themselves?

The Chapter at a Glance: Take the Chronometer Test

The next time you look at your watch, ask yourself this: If misunderstanding is the rule in communication, how should I express myself? How clearly, precisely and concisely do I express myself, already? How could I be more precise, clear and concise? Do I articulate my expectations? Every time? Do I always double-check the expectations of others?

Manage Faster Than You Can Think!

"Slow down, I'm in a hurry."

German Chancellor Adenauer to his chauffeur

What Was the Boss Thinking?!?

Have you ever asked yourself this question? What answer did you give yourself? Let me guess: "Nothing!" At the end of the day, the boss is a manager. And we all know: managers aren't paid to think, they're paid to perform. What about you?

Suppose an employee with a management role comes into your office and informs you about a serious problem that has suddenly arisen. What do you do? "I don't do anything at first", a guy at a seminar once said spontaneously. The other participants laughed out loud. The laughter died down when they saw me applaud: this was the right answer.

Managers are people of action! And that creates problems!

And because it creates problems, it is usually better to do nothing, at first. This beautiful management paradox is demonstrated perfectly in the following example:

Kevin S. is the CEO of a large Western European corporation and currently at a leadership conference.

The top managers gathered together hang, breathlessly, on the words of the sophisticated analysts and their stories of success, including easyJet, Puma, Google and Virgin Airlines. The atmosphere in the conference hall is euphoric. Everything seems so easy: copy the recipes for success and you'll go down in the annals of the great.

Kevin is excited about the new ideas and thinks: "That's how we're going to do it too!" While the speaker is still on stage, he takes out his iPhone and shoots off an email to his right hand man, Manuel: "Great new approach: We reorganize our sales activities, concentrate 100 percent on our core business and a few target groups. Administrative ballast, unproductive activities and non-profitable segments have to go—our profits will go through the roof! We'll be as profitable as easyJet. Manuel, please get started immediately on the plans. Thanks. See you soon, Kevin."

Manuel, boss of the strategic planning team, stares at the email for several minutes, beside himself with anger and disbelief. He later complains to me: "We just started to rework our strategy last year and began a reorganization of the entire corporation. The results are better than expected. Why on earth would he want to throw it all away, now? We can't possibly copy easyJet! That's a completely different world. They're a cheap airline with undemanding customers. We have high tech products and customers that are happy to pay a premium for that. What was he thinking?" Nothing. He didn't think about it. He just did it.

Manuel is caught, cursing and swearing, between a rock and a hard place: "What should I do now? As crazy as the idea is, I can't ignore a direct order from my boss!" So he grits his teeth and familiarizes himself with the idea. He studies the business model of easyJet and figures out a way to integrate these ideas without doing damage to the new corporate strategy. He works until midnight, two nights in a row, on a presentation to show Kevin.

When Kevin gets back from the conference, he gets Manuel on the phone immediately for a meeting. But before Manuel can turn on his laptop, Kevin says: "Listen, Manuel, that idea with the change of strategy... that was just one of many ideas that went through my mind at that fantastic event. Let's put that idea on the back burner for the time being. Things are going so well for us at the moment, why change strategy? But, next time you definitely have to be there! You'll come away with ideas for the next ten years! And? How was everything while I was away?" Manuel just stares at him. He picks up his laptop and leaves the office, making up an excuse on the way out.

He charges into the men's room and stands next to a now very startled financial director at the urinals and bursts out with: "Why don't people think before they press the 'send' button? Two days of my life, wasted on his brain fart! Two days!" He punches his fist against the wall before going right back out again. The Financial Director almost wets himself from shock. He didn't have a clue what Manuel was talking about...

What happened to Manuel is hard to understand: How could Kevin treat his number one gun with so little respect? Did he think that Manuel had a 48 hour day? No, he just didn't think before he pressed the "send" button. He was a victim of the Lucky Luke syndrome.

The Lucky Luke Syndrome

Lucky Luke was the cowboy who could draw a pistol quicker than his own shadow. Kevin is the boss who manages quicker than he can think. That's not bad, it's normal. It's the occupational illness of the powerful.

> When you're always under pressure to perform, forgetting to think first becomes a way of life.

If you're feeling another "mea culpa" rising in your throat, choke it back for a moment. Of course, we can all relate to Manuel's frustration. But before we beat Kevin, the thoughtless quick shooter, over the head with a 4 iron, let's ask ourselves the essential question: Is thoughtlessness in management really a sin? Let me ask it another way: "Would you want to have Kevin as a boss? The answer can only be a resounding: "Yes!" Because Kevin was the first one in his industry to introduce value added services. He was the first one to offer online support. He was the very first in many innovations because he regularly jumped in with both feet while "smarter" guys were only thinking about it.

The West was won by the brave, not the clever.

The successful ones are the ones who get moving, while others are still analyzing. We all, myself included, curse the sales managers who unthinkingly sell products that haven't yet been invented—thereby pushing developers and production managers to the point of burnout, to invent and produce what "that idiot" came up with again. But if we're honest: These "fast guns" bring in the bucks with which we pay the moaners and groaners.

I once worked with a head salesperson who sold faster than anyone could think. She sold, literally, everything, without giving it a second thought: whether the products existed or not, the moon, the stars, her grandmother and what the Germans like to call the egg-laying, milk-producing, wool bearing pig. The result was chaos, a lot of confusion and a heap of mistakes. But also a ton of new customers who were very excited about our incredible innovations. This was how we managed to become one of the first banks in our country to offer loans in foreign currencies, though we couldn't handle it, administratively, at the beginning. While the customers were happily signing their loan papers in the front office, we were in the back slaving away over hasty exchange rate calculations, creating account statements and other spreadsheets, manually. A crazy makeshift solution! We cursed our head saleswoman and her "thoughtlessness" during a period of several sleepless nights. But during the day, we thanked her for all the new customers. What I'm trying to say is:

People who act without thinking are like basketball players who can't play defense. They make most of their shots, but lose the game in the back court.

In plain English: Management needs fast guns (above all in bureaucratic, anti-entrepreneurial organizations). However, fast guns don't just need a waist high holster with a loaded Smith & Wesson. They also need to be able to take their fingers from the trigger now and again—and press the pause button.

Press the Pause Button

Kevin didn't have a pause button. That's why he pressed "send" on his iPhone without considering for a moment what the consequences could be. A few weeks after this incident, we tried to create a pause button for Kevin in an informal mediation talk. "Maybe you could just sleep it over one night before pressing the 'send' button next time?" I recommended.

"I can't do that when I'm excited about something", admitted Kevin. Then he turned to Manuel and said: "But how about next time, instead of sitting down for two days to work on one of my ideas, just think about it for 10 minutes and make two columns: Put the advantages of my idea on the left side, and the obstacles and risks on the right. Then we'll decide together whether it is a go or a no-go." Manuel was visibly relieved. His trigger-happy boss was finally doing something about his thoughtlessness.

Stay (or become) a quick-shooter! But give yourself a (mini) break to think before you let loose.

Or, as the saying goes: "Look before you leap." Where should you be looking? At the consequences, and by that I mean …

- ... the **collateral damage** consequences: When you set A in motion, what happens to B, C, D ... ? What's the overall result?

- ... the **communication** consequences: How will you communicate the new ideas? How will the people on the receiving end react to your actions? How will colleagues, employees and customers react?

- ... the **financial** consequences: What will it all cost? Will there be any opportunity costs? Where will the budget come from?

- ... the **capacity** consequences: do you have the manpower and all the other necessary resources?

- ... the **big-picture** consequences: What do you stand to gain? What will the ROI be? Or for Kevin, the ROS—Return on Send?

- ... and finally, the **personal** consequences: Do you have the time and energy to coach and support the plans once they've been set in motion?

This is not going to take that much time and effort. An experienced manager can evaluate the consequences of the quick-shots within a few seconds—Kevin as well, now. Since he regularly asks himself what consequences his flashes of brilliance will have after our intervention. He even presses the pause button sometimes. He minimizes negative consequences by communicating to others or, for example, ensuring that Manuel only invests ten minutes instead of two nights. This new found reflectiveness is a huge relief for his employees. Kevin is doing himself the biggest favor of all, though, because speed junkies who never push the pause button, do themselves the most damage: they lose the respect of others.

No Respect for Speed Junkies

When I was visiting a company in the telecommunications industry, I was surprised at how relaxed the employees were about the latest management board announcement that the company was going to be completely reorganized. One employee explained to me: "That's our new CEO. The old one decentralized everything. The new one will centralize everything again and that will probably take about six years. Since the new boss will most likely only be here for four years, like the last one, we'll just wait it out. The next CEO will put everything back exactly the way it was before, to avoid having the same strategy as his predecessor." How much respect do you think these employees have for their new CEO?

If you work faster than you can think, you will be stigmatized, ignored or blown out of the water.

A hardware developer for security systems once told me: "My boss is a speed junky. Every day he gives me at least two new "ground-breaking" projects which are, supposedly, super urgent and have to be conceptually designed within the week. I normally leave them sitting in my inbox. Only after he's asked for it a third time, do I get started." The look on my face must have betrayed my discomfort (you don't normally expect that kind of insubordination from a German engineer). He put me out of my misery, saying "Only 20 percent of his ideas get asked about a third time. All the other projects he forgot the minute he went out the door. Why should I follow through on the ideas of a Boss with 'Management Alzheimer's'?" How seriously do you think this hardware-developer, who is excellent by the way, is taking his boss?

If you shoot faster than you can think, you won't be respected.

That's bad enough on its own—forget about the economic damage it could be causing.

Mini Test: Are You too Quick?

- Do you call your employees during their vacation to tell them about your great new ideas?
- Do you call your employees when you're on vacation to make sure you're not missing anything?
- Can you turn off your cell phone for an entire hour during the day?
- Which employees have you overwhelmed with your quick-fire ideas this week?
- Do your employees back you and your ideas? Or did they give up on you a long time ago?

The Machine Gun Boss

I know a company president who overwhelms her employees by running around so fast that they can't keep up with her. Out-of-breath and over-heated, her employees fall around her like flies close to a light bulb. She simply wears them out with her supersonic tempo (which she expects from them as well). She bombards them with tasks, projects, challenges, and innovations, as if she were spraying them with quick fire bullets. But, have they really understood a single thing? Can they see the big picture? Does the company have the capacity for it?

Her employees are running around so fast, they don't have time to reflect on what direction their machine gun president is taking them. That's frustrating, because nothing is more demotivating then disorientation. As a compensation for the stress and unremittingly fast pace, the board pays extremely high salaries. Which doesn't really help.

Her best people are under such pressure, they all leave at the first call from a headhunter, in spite of the high salary. One of them admitted to me: “No one can keep up that pace for more than a year! You've got to get out before you totally burn out!” The employee turnover at that company is unbelievable, and we all know what turnover costs. The president remains tough: “But we have to put the pressure on, otherwise we'll lose our place at the top!” It's her little niece, of all people, who finally helps her to see some sense. When she goes to one of her soccer games.

There, she sits next to a trainer in the local girl's league who had once played professionally. “Our coach back then pushed us so hard in training we were as fast as the best marathon runners and decathlon athletes. We just lost all of our soccer games because we were so tired from training—we could barely get our feet off the ground.” He looked at the company president and said with hard-earned conviction: “You can't win if you wear your team out.”

The president is too ingrained in her ways to change herself, now. But because she's a woman of action, she takes things into her own hands and makes some changes. She hires an assistant who is trained in coaching and process management and positions her between herself and her worn out team. The assistant is responsible for shifting things down a gear or two when necessary. She makes sure everyone sees the big picture and acts as a shield when the new ideas of the overactive president go flying. She organizes kick-off meetings, workshops and other process supporting measures. Personnel fluctuation drops by half. Productivity goes up as well.

Take a Deep Breath, Then Press Pause

Peter M is the new head of sales at a plastics company. He knows that the first hundred days in the job are crucial because the board's plans are extremely demanding. What does he do? He gets started. He cranks out a new project every day. He's like a human leaf blower, blasting in innovations. After a few weeks, a female regional manager says to him in a meeting: “I think we’re overwhelming our customers and sales people with the new campaign. They still haven’t digested the last ones.” This warning isn’t the first, but Peter isn’t open to it. After the meeting he complains: “That woman wants to put the brakes on everything. Who does she think she is? We’re tearing our hearts out here and she’s just dragging us down. It’s time to roll our sleeves up, not sit around being so negative.” Peter is a typical machine gun boss who shoots first and then asks questions.

Luckily, after taking a deep breath, Peter recognizes: “When I’m going too fast I just do myself and the company damage—not to mention the employees.” That’s why he forces himself to take a short break before important decisions to think things through. All excellent high performers follow such a recipe for success. I have collected a few over the years (by the way, thanks to everyone out there for the great ideas) and I’ve listed a few of them below—anonymously, of course:

- “Before I do something, I consciously take a short break to think. Even if it’s just for a few seconds—it always helps!”
- “Before I give the go ahead or press the send button, I breathe in deeply a few times and ask myself: ‘Do I really want to send that out now?’ If I can’t think of any good reasons not to, I go ahead.”
- “When I’m assigning a task that will take up more than one person-day, I prescribe myself a three-hour break from making the final decision. During this time I take care of normal business and if I still think the decision is a good one, I finalize it.”
- “When I’m excited about something or see that something needs to be done, I just have to act. Immediately. But I always ask myself at the same time: ‘How will this come across to the people I may be overwhelming?’ People will accept my idea much more easily if I talk with them about it first.”

- "I know how impulsive I can be. Before important decisions, I always go to an old board member colleague. She's the opposite of me: she loves to put the brakes on: I say to her 'Quick. I need two minutes of your time. Give me every reason you can think of, why I shouldn't do this.' If she manages to give me some good reasons, I shift down a gear. If she can't, then I know: 'Go for it!'"

- "I'm aware that I am often too quick for my employees, but I've found the best remedy: planning. I don't make demands anymore like 'Get it done, now!' Instead I say: 'Plan it out. You've got two hours. Get the rough costs, time expenditure and objectives down on paper and then we'll talk about it.' The results clearly show which ideas we should go with and which ones we should toss."

- "I am, and always will be, a man of action. I develop the ideas, visions, and strategies. But I've set up a hand-picked group of employees that I bombard with my ideas and they have twenty four hours to tell me what they honestly think."

- "I've gotten mindless, non-stop running around out of my system: whenever I begin to feel the urge, I ask myself: "When I do it right away, what am I trying to achieve? How will I come across in the end, competent or incompetent?" That helps and I go forward in a much more level-headed way."

- "I ask myself the decisive question: 'Okay, we *could* do this now—but do we really *have* to do it?'"

- "I ask myself: 'That is a good idea, but do I really want to burden my employees with something extra?' And if yes: 'How will I reduce their workload to compensate?'"

- "I've modified the well-known Walt Disney technique. Whenever I have a good idea on my 'idea seat', I sit down on the 'leader seat' and ask: 'How will I present this to my staff, now?' Then I sit on the 'finance seat' and consider the costs: 'How much will it cost? What will be the return on investment?'"

- "Don't laugh, but I put on my thinking cap in the restroom. It sounds a bit crazy, but it helps. And, it's so easy. I've got a nice little retreat where I can have a moment alone and consider the consequences before I run with something that could bounce back and hit me like a boomerang."

Did you find something there that will work for you? What? And if not: What recipe for success would you create for yourself?

It doesn't matter what technique you use or how you use it, make sure you add some intelligence to your mindless, non-stop activity.

It's a bit like your gut feeling. It may be a great advisor, but if it's not combined with numbers, data, and facts, it leads inevitably to kicking the spectacular goals for the other team (unless we're talking about a matter of life or death, in which case go with your gut).

By the way: Peter is happy that he didn't fire his "negative' regional manager that day. Instead he thought it through after taking a deep breath and pausing to think for a few minutes. He's gotten used to her critical style, now, and he's accepted it. She is the best of his three regional managers. If he had fired her without thinking about it first, he would have done himself, and the company, a great deal of damage.

Managing Via Highlighter Pen

The following story actually happened. Two divisional heads were reading a new management bestseller with great appreciation. They marked all the passages they thought were the most important with a highlighter pen and passed their copies on to their next in line, asking them to implement the new ideas.

"What?!", you say, "If leadership is that easy, why do we need managers at all?" Throw out all the executives and hire lecturers, editors and writers! They can read books and highlight texts the best. And take away the expensive cell phones, tablets, notebooks and company cars! Because all that a manager needs is a highlighter pen. And employees, who can implement the highlighted texts ...

The employees who had received the highlighted management texts knew what to do. They took the books, said "thanks" and fobbed off all subsequent questions about their progress with: "Yeah, the implementation is going great! Couldn't be better!" If managers are naive enough to think that transformational projects can be achieved without feedback, cover, commitment, engagement, resources, budget and supervision from above—meaning without leadership at all!—then they will also believe the employees who lie and say that the implementation has been successful. In fact, the aforementioned employees didn't even read the books. Why should they? They knew right from the start that the two divisional managers couldn't really be serious.

You can't manage employees by bombarding them with ideas.

When I was still a young apprentice, my very first boss said to my parents: "Now, Klaus has hundreds of ideas, but he doesn't finish any of them." That was tough to hear. But good managers can not only take "tough love", they can turn it into something advantageous: I swore to myself back then to convince my boss of the opposite.

It's still a big effort for me today but it's gotten a lot easier: I do one thing at a time until it's finished, instead of jumping to the next idea while my employees are still busy with the last one. I am (still) a man of action. But I have begun to think before—and during—my actions.

Stay a man or woman of action, but for God's sake start thinking things over.

If you bring up a new idea, you are also obliged to think it over. You might want to consider some coaching, particularly for projects where thinking things over from start to finish will take more than 5 minutes.

Über-performers

Why do some managers manage faster than they (can) think? They are often driven by two emotions: fear and/or success addiction. The company president's statement from the example above is very revealing: "We have to put the pressure on, otherwise we won't stay on top!" She was obviously afraid of losing her place at the top, it wouldn't have helped, or changed anything, to advise her: "Don't be in such a hurry to get things done! Take some time and sleep on it, first!" This clever advice is considered trivial by public speakers and management book readers alike: such tips don't help. Fear is always stronger than reason.

It doesn't help to fight fear. That only makes it stronger. It's better to give in to it fully, take it seriously and think it through to the end.

That's what the company president did in coaching. At some point she said: "When I am afraid that we will lose our position at the top, and when I put my people under pressure again, and make them want to leave the company in droves, that's when I really do have a good reason to be afraid. Because I will definitely lose our position at the top from lack of personnel!" This simple insight has provided her with a measure of balance that enables her to continue being a woman of action. If she forgets to reign herself in, a brief reminder from her assistant is enough to get her back on track.

Fear makes you blind—and not very successful. Just like a success addiction.

Most über-performers are addicted to success. They want to get things started, make a difference, initiate something and get ahead—come what may and by any means necessary. It can't happen fast enough. That's why they whirl around like a vortex, with their employees left standing in the dust. They're unconsciously slamming the brakes on their high-flying goals, because without the help of their employees they can't achieve a thing.

If you're running around too quickly, you'll slow yourself down. Less is often more!

You all know the meaning of the saying "haste makes waste".

When I advise über-performers to take some time to think things over, some reply: "You know, you can analyze things to death!" Of course, I recognize such mulish answers for what they are: the denial of a speed junky who doesn't want to admit that his or her whirlwind activities won't lead to success.

Doing and thinking are not diametrically opposed.

Think of it more as a diametrical extreme along a continuum whose truth lies in the proverbial medium: a manager who only pushes forward—too quickly—does as much damage as someone who is grumbling, whining and lazy. Extremes are never good. What we need are high performers who can find the right balance.

Balanced High Performers

Take a look at your own area of responsibility: What's your team like? Are they thinkers or über-performers? On a scale of 0 (= only thinking, analyzing, planning, conceptualizing) to 10 (= only action, without a lot of thought)—where would you put yourself on the scale, your employees and your team as a whole? I have been able to achieve a 50 percent increase in productivity in consulting projects just by asking this one simple question. Because line managers normally give "off the cuff" answers like: "We analyze and worry a lot but don't get much done!" Or "We go off like loose cannons without a plan or a concept!"

Balance your actions with your thoughts.

How can you find your equilibrium? Beyond pretense, beyond affectation. Imbalance is driven by emotions: An über-performer is afraid of reflection or of standing still and missing the boat. They want to avoid failure at all cost. Thinkers fear action, embarrassment, mistakes, change and being overwhelmed. The great thing is: We all feel it when we, or our system, is not in balance.

As soon as you become aware of imbalance, take care of the emotion that's causing it.

When they become more aware, high performers start listening to their gut feelings. I know a board member, who has made it a part of her HR development approach to go to meetings where no results are being achieved. When the participants start going around in circles, she often says things like: "Why are you still beating around the bush? Are you afraid that I will tear someone's head off if the project fails? I'm not going to do that. The worst case scenario is that we have to allocate the costs to our training budget." Often the pointless discussions end within seconds and the managers start concentrating on new tasks: They get back to high performance (finally).

The aforementioned board member also visited departments and project teams who were known for their aimless activity. Here, she liked to hit the nail on the head, once saying: "Wait, the meeting isn't over. Everyone stay put and explain to me how you plan to get this big project started without an action plan, without budget planning or any kind of a schedule. You're all worried that this will turn out to be nothing more than a pipe dream if you look a little closer, right? You don't have to worry. I promise you that we're going to transform this idea into a reality, because it's a good one. But if you can't give me at least a rough draft of your plans, you won't see a penny of profits." The people sat down grumbling—but they started to think (finally).

Never lose sight of the balance between your thoughts and your actions—in your own area of responsibility and within the team as a whole.

That's not only true for business, it's also true, as my clients and coachees assure me time and time again, for all the other areas of your life—in your personal relationships and family, when raising your kids and in conflict situations. If you find the right balance between your thoughts and your actions you will not only be more successful, you will also go through life more relaxed and more satisfied with yourself and the world.

The Chapter at a Glance: Take the Printer Test!

The next time you print out a document (or ask someone else to), ask yourself while you're waiting: Am I managing faster than I can think and expecting my out-of-breath employees to keep up with me? Do I consider the consequences of my actions? How can I get action and awareness back in balance?

Don't Trust the Opposite Sex!

"I would rather trust a woman's instinct than a man's reason."

By Stanley Baldwin

Who Make Better Managers—Men or Women?

That was the title of an article that I put on my homepage a while back. Guess which one of my management articles is clicked on the most? When my IT support told me the click rate, I must have looked as if someone had told me that all politicians had become honest overnight.

As a modern, enlightened man, who's grown up in the era of equal rights, I had thought that the exasperating gender discussions had finally come to an end. Which just shows what kind of ridiculous nonsense modern, enlightened men are able to come up with. The truth is: the gender fight is alive and kicking. Maybe it's not as obvious as before, but in its modern subtlety it is still on the rampage.

Minutes after I had placed the somewhat provocative question in the internet (were men or women the better managers), umpteen million (slight exaggeration) clicks attested to the dogged speculation on both sides of the gender divide. Both hoping to witness the shamefaced retreat of the other, and enjoy the triumphal validation of their own sex. Are we in kindergarten, here? How primitive is that! About as primitive as it is in the workplace. Two words: glass ceiling.

It's more than perceptible in many companies: women only climb to a certain hierarchical level, then it's "game over" when it comes to more climbing. It doesn't matter how talented and qualified the woman is, she can have a PhD and two Nobel prizes to her credit—she will usually never make it past divisional head. "A woman on our board?" asked a female divisional manager of a retail company, rhetorically, "You're more likely to find a bar of gold in a trash can."

Another example: the old boys' network. The name, alone, implies the discreet charm of colonial chauvinism, because there is no old girls' network (there should be). If board members lie, cheat, ruin companies, evade taxes or get tarred and feathered on their way out, they still manage to pop up, magically, afterwards in some other board room in another corporation a short time later. The old boys' network took them back into the fold.

When female managers fall, they usually fall into a black hole, disappear into thin air, and at best manage a hot dog stand somewhere on Coney Island. No net catches them up in their fall from grace. Worse yet: when women fall from the career ladder, the press circle in like vultures to scoff and jeer—original German headline: "Are women not yet ready for business?" Translated: men shoot women down from the ladder, then mock them on the way down, even label them as immature. Is that why so many managers clicked on my provocative article in the internet? That would seem the obvious assumption. But I think there's a different motive at work, here:

Even the most primitive chauvinists think discrimination stinks.

Not because they've suddenly become woman's best friend—but because they've finally begun to see that their years of taking pot shots at women have really been an exercise in digging their own graves.

Digging Your Own Grave ...

Many male managers are experiencing increasing feelings of guilt in the continuing battle of the sexes. An acquaintance of mine experienced two events in a single day. Both of which gave him an uncomfortable pause for thought. One was the recruitment process for a vacant position at his company: five applicants were up for selection, four of them were women. Who do you think got the job? The only man in the group. The other was a project meeting which took place a couple hours later: a young, female project manager had been taking care of a major customer while at the same time working on a follow-up project for them. She knew the customer and the new project inside and out. Who was given the job of project manager for the new project? Not her. It went to a male colleague who had only been involved as an engineer up until then.

"It's ridiculous. I think I made two huge mistakes today", said my acquaintance. But that was an understatement. In my opinion it was nothing less than company sabotage.

When the customer heard that "their" project manager would not be responsible for the project anymore (instead it would be the engineer, one they had worked with on an earlier project and considered, in polite English, to be an "arrogant a******') they expressed a sudden burning desire to go to one of the competitors. Impending damage: 100,000 dollars.

Why did this acquaintance of mine, usually, a very normal, intelligent person, take a project away from a woman who had proven her project management competency beyond a shadow of a doubt? Why, instead, did he give it to a man whom even the most agreeable colleagues would not have considered their first choice? Why couldn't he trust the opposite sex?

Why Don't We Trust the Opposite Sex?

When a Chinese firm took over a German company, recently, the Chinese CEO went from department to department to talk with the employees. The poor Germans, not used to this from their own line managers, were a bit overwhelmed by this sudden attention. They asked themselves skeptically, "Why's he doing that?" The office grapevine provided the answers: "He's listening to us! So he can decide who to fire. They're going to start laying people off." What do you think happened next?

Right: The Germans closed ranks—they gave the Chinese CEO the cold shoulder. He was at a complete loss. "In China it's normal to visit the staff before starting to work together. This allows everyone to introduce themselves and to talk a bit about non-business things. Strange that this kind of politeness doesn't exist in Germany", thought the new CEO. But the Germans weren't behaving like this because they wanted to be impolite, they just didn't have a clue about the Chinese culture. The battle of the sexes is like a battle between different cultures.

And what people don't know, people treat with distrust.

Such as: "I have no idea how the old girl ticks, but to be on the safe side, I won't believe in a thing she says." Of course, this is an anachronistic, atavistic, paranoid, Neanderthal reaction. But some men can be anachronistic, atavistic, paranoid, football watching, car loving, beer drinking, burping, snoring, thigh slapping Neanderthals—some women too, by the way. Because, years ago, when I wanted to free myself of my Neanderthal ways, it was a woman who slammed the door shut on that.

I had just begun reading the book "Fish without a Bicycle" by Elizabeth Dunkel. The book was lying on my desk when a female friend, who happened to be visiting me, grabbed it off my desk and said with a deeply moralistic tone of voice: "Excuse me, but this book is for women. Why are you reading it?"—as if she had caught me with the newest Victoria's Secret catalog. "Well", I answered a bit naively, "I work with a lot of women and I want to understand how they tick." I shouldn't have said that. It can be risky to admit to wanting to understand the opposite sex. That would be admitting that they're different in some way ...

My acquaintance didn't understand me. She didn't think it was right that I should feel the need to "understand" women. She met my desire for more understanding with misunderstanding.

The battle of the sexes is being carried out by both sides. Not out of conviction but from a refusal to understand.

My acquaintance was in danger of losing an order of 100,000 dollars, not because he hated women or because he would like to shoot them off the ladder too, but out of pure ignorance—because, unlike me, he didn't read books about or by women. That's the big problem, but at the same time the solution.

Men and Women are like Foreign Languages

Meaning, it's possible to learn. If my acquaintance had known more about women, he wouldn't have attempted to replace the female project manager with a man.

The end to the battle of the sexes is not peace, but understanding.

I witnessed a tentative endeavor to develop more understanding in the company kitchen of a client.

> He: "Um, Stefanie, excuse me for asking such a stupid question. But why does Monica always come to work so made up. Is she interested in someone? I thought she was married?"
>
> She: "Monica doesn't dress like that to impress men, Michael. She just enjoys wearing make up and designer suits that show off her figure."

Her colleague opened his eyes. As he passed on his groundbreaking insight to other colleagues, the attempts to hit on Monica, which were killing productivity and driving Monica to her wits end, came to an abrupt halt in a heartbeat. Some of the men asked themselves the alarming question: "How will we know if a female colleague is looking for love, if not by the depth of her décolleté?"

Understanding for the opposite sex doesn't come on its own. You have to go look for it.

This can be arduous but often refreshingly satisfying. In addition to a much better working environment, you'd also make a mountain of money. For example: 100,000 dollars. That's how much my acquaintance made for his company when he decided to approach the female project manager face-to-face and replace his doubts and prejudices with newfound understanding: he transferred the project back over to her. That was surely not an easy discussion. It never is, as I learned at the innocent, young age of sixteen.

Desperately Seeking Understanding

One of my first bosses was a young mother with a full-time management position. And so it came to be that I, a promising young man and apprentice, was given the lofty, portentous, enormously responsible task of picking up both her daughters from kindergarten every day at noon. I would have rather cut my finger nails with a buzz saw. Have you any idea how tortuous it is for a sixteen year-old boy to stand outside a kindergarten every day, waiting for the little sweeties to emerge meanwhile suffering the astonished looks and giggling gossip of the mommies gathered around? It was humiliating, especially for a young man in Austria, back then, I can tell you.

Today I know that my boss was way ahead of her time. Long before work-life balance had become one of the trendiest phrases of modern times, my boss was living the perfect mixture of work and family life. What the modern working world has difficulties with today, she took for granted back then: work and family go together. You can't split a person into "private" and "business", without driving them into an early burn-out.

It wasn't easy for me to learn this lesson from my boss. Understanding others almost always hurts, at first, because it usually goes against one's own prejudices and dogma. But where understanding flourishes, the battle field disappears. Isn't it nice when men can learn from women? Wouldn't it be great, if women would return the compliment?

OK, that wasn't really fair as the following example shows: I overheard a male chemist say to his female colleague at a break during a meeting: "You know, Barbara, I've learned a lot from you, for example, that you should listen to the opinions of others before making a decision. But you just did something kind of dumb back there. If we're talking about what kind of Bunsen burner we should buy for the trainees, you don't need to ask for everyone's opinion! If you aren't comfortable making a purchase for 500 dollars on your own, that will be seen as a weakness by the others."

I was just waiting for a defensive reaction like: "But you can't make decisions over the heads of the others. Consensus blah blah blah, harmony, blah blah blah, macho behavior blah blah blah." Instead the Ph.D. level chemist raised her eyebrows and said: "Oh shit, you're right!" And then both dived into a five minute discussion about which types of decisions needed consensus and which, definitely, not. I'd be willing to bet that this Ph.D. manager is now seen by her colleagues as a very decisive and confident leader, because she had been able to learn something from a—God forbid!—man.

> We have to stop competing with each other and instead learn from one another.

And it's because there are so many women with careers in business that it is a management sin of the first order to waste all this potential and not use the synergy of both sexes. "Cooperation instead of confrontation!" that's what it's all about (and it's not just reserved for the battle of the sexes).

Xenophobia and the Damage it Causes

Let's take a look back at the chemistry Ph.D.. What probably occurred to you: the roles could just as easily have been reversed! I know a lot of men in mid-management who have to ask permission from their directors before they can buy a few pencils. You can imagine what colleagues would say about that: "He's gone to ask the boss if he can blow his nose." Or: "He has to arrange a board meeting to take a piss." Think that's funny?

It's not just women who are discriminated against in the workplace (and elsewhere).

We fight against anyone who seems different from us. Mick Jagger made this sentiment immortal in his song "Satisfaction": "He can't be a man, 'cause he doesn't smoke the same cigarettes as me". The accruing costs of xenophobia (a fear of what's strange or foreign) leave the skyrocketing prices of oil in the dust.

My previously mentioned friend almost lost 100,000 dollars because he excluded a woman in a knee jerk reaction—probably women, in general, are simply foreign to him. How many billions are lost in business each year because we unconsciously discriminate against colleagues who seem different, instead of taking advantage of their abilities more cleverly? How much is your company losing, regularly, because of such behavior? How is your area of management being affected by this? Every day, horrendous sums are lost due to this kind of xenophobia. In every failed project team there is (at least) one person who says afterwards: "I could have told you from the beginning." And no one listened to him or her because we prefer to ignore the minority voice rather than integrate it consciously. The maxim is therefore:

Go for diversity and integration, not uniformity and isolation!

If you manage that, your teams and departments will make quantum leaps in increased productivity.

The Power of Integration

The power that integration can generate is reflected in this statement by a divisional head at a telecommunications company: "We had around a hundred and forty teams spread out in different departments that are completely isolated from each other. Birds of a feather had flocked together. Over the years those interested in innovation or customer orientation managed to end up together in their own cliques. Curiously, we had many all-men teams and many women-only teams. Even the smokers stuck together."

What sounds kind of funny was actually a disaster for this division's productivity. If teams are no longer measured by the sum of the individual abilities required for a specific task, but based on primary gender organs or the preference for filter-tips, productivity and turnover will take a nose dive. Because the innovative team lacked someone who had their eye on customer needs—so that great new inventions were made with blissful ignorance that would never make it in any market. And the customer orientation team was treading water without a single, innovative impulse ... and so on.

The divisional head reported further: "Teams are not successful when likeminded people slide into a rut of conformity. If that were true, the gangs of New York would have taken over the city long ago. Teams are productive when they manage to integrate as many different capabilities as possible. That's why we broke up the hundred and forty teams and built new ones—in such a way that they have the maximum amount of variety. Now that we have groups where the employees are integrated according to their abilities and with an equal distribution of men and women, I hardly recognize our company anymore!"

Productivity has increased above all expectations, costs have mysteriously dropped. Our turnover in relation to capacity has increased above average. The director had to admit: "I always thought that profit only improved when employees were let go or new markets were conquered." That it's possible to be successful by correctly integrating employees according to their abilities is obvious now.

Sure, some will say: when men and women, innovators and craftsmen, engineers and customer service personnel are suddenly let loose on each other after years of comfortable isolation, the attack of the Huns on the Roman Empire will seem like child's play in comparison. That's right. That's why it's called team integration. It's a (development) process, which requires professional mentors and coaches. In all, there were twenty professionally trained team coaches who facilitated the "Attack of the Huns" on the individual teams to ensure that the enormous energy being set free, broke new ground constructively. Nothing begets nothing. If you want teams able to perform, you have to be able to lead them.

What abilities have you integrated into your projects, today? Where do you notice a tendency towards isolation? Where do you tend to isolate yourself? And how can you become a great integrator? It must be obvious, by now: diversity managers are the most successful by far, and they are also the most satisfied.

To quote my old acquaintance once again: "It is very tiring to constantly cut certain people out. I feel a lot better since I don't do it anymore."

The Chapter at a Glance: Take the Diversity Test!

The next time you see someone that seems too different or strange at work, ask yourself: what bothers me about him or her? How can I learn to understand why (or for what reason) he or she is doing or saying (or not doing or saying) something? How can I (better) integrate diversity into the team? What can I learn from others?

Be an Unscrupulous Scoundrel!

"Nobel be man, helpful and good. For that alone sets him apart."

Johann Wolfgang von Goethe

(Im)Moral Management

Corruption, bribery, extortion, executive salaries ... The greatest financial crisis since the invention of capitalism—the subprime mortgage crisis—was not caused by shortages, inflation or natural catastrophes. The crisis was entirely man-made and basically the result of bankers who, in an orgy of excessive greed, gave out loans that no reasonable person would have ever approved. Why did they do it? Because they were being paid on a commission basis and they couldn't pad their bank accounts fast enough. "Morality", for them, was just the name of a play on Broadway.

A former car salesman once told me: "When the banks on Wall Street paid out their annual bonuses, the local Maserati and Ferrari dealerships couldn't keep up with the luxury car orders." Politicians can be bribed and managers are corrupt—that's what most people believe today. But who cares what the average person thinks as long as I'm getting the biggest piece of the pie, right?

But, of course, you don't think like that. Managers who read books rarely take bribes. Corruption goes hand-in-hand with a certain lack of education. You've got to be pretty thick-skulled to believe you can compensate for intellectual deficiencies with money. Or to put a twist on a John Candy quote, "If you're not a man without millions, you won't be one *with* millions, either." Apropos bribes: Have you ever taken one?

Have you Ever Taken a Bribe?

I know, that's not a question you should ever ask a top manager. Because to get to a certain hierarchical level, you'll almost always be faced with the temptation either to take one or to give one. The machine that business professors naively call capitalism doesn't function any other way.

Have you ever given in to temptation (or not?) and regretted it? Managers don't usually talk about corruption. Sometimes it bursts out, though—in coaching sessions, for example. A while ago a German corporate crisis manager came to confide in me. He was supposed to save an international subsidiary from ruin and simultaneously achieve a substantial turnaround for them. If successful, he was guaranteed a position on the board. A failure would mean the end of his top management career at this company. So the troubleshooter gave it his all, laid off as many as possible, cut costs, reorganized every nook and cranny. But the increase in turnover that the board was expecting would not be possible in the allotted time frame. Until the day came when a large customer winked his way with a more than tidy order amount—and held the other hand out demanding a six figure pay out.

What was truly surprising, here, was that the crisis manager didn't pay up. He felt a real moral dilemma. He sat in coaching, looked at me with furrowed brow and asked: "Should I pay? Should I stay honest? What would you advise me to do?" His marketing director was pressuring him to pay, his financial director assured him some creative accounting would keep him out of hot water, the pressure from the board was mounting every day.

That's what corruption looks like in the real world. It's not (predominantly) about greedy self-enrichment, as some columnists like to report, although they have never worked a single day of their lives in a real corporation. Corruption is not a case of mad money grubbing for most managers, but intense psychological "dis-ease" or in plain

English: (emotional) blackmail. Either you pay or you suffer the consequences. If you don't pay, lots of people could be out of their jobs, possibly even you. If you do pay, you secure jobs, you're a hero and everyone applauds you—until the whistle gets blown. Then the exact, same people will be kicking you out on your backside, whose jobs and executive positions you saved with your "donation". Do managers have to put up with that kind of corruption?

Don't Put up with Corruption!

Let's start with the small stuff. When I came to Serbia a few years ago, it was a country in the midst of sudden upheaval. Patrolmen considered foreigners easy game. You got stopped at every corner and charged with some misdemeanor that suddenly sprang into the mind of that street's preserver of the peace. I don't carry a grudge against any of them, today: those people were not exactly swimming in luxury. One of them even complained to me: "Should a policeman not drink coffee, just because he can't afford it?" I always took plenty of t-shirts, pens and other giveaways with me on those trips ... and I would like to meet the nitpicker with the nerve to call that "corruption". It takes on a totally different aspect, though, when we're talking about large sums of money. The problem is not that some managers are corrupt. The problem is that the others don't know how to deal with it.

If a kidnapper calls and says: "We've got your wife, give us a million or she's dead", only a few husbands would fall into the quagmire of this proposed dilemma. The options offered are both completely out of the question, at least when considered in the light of day, because there is a third option: inform the police, immediately! When we're talking about corruption, though, only a few managers have been able to come up with the third (or fourth, or fifth) option.

One option, for example, is networking. I know a salesman in a very competitive and corrupt (to the bone) industry, in Europe, where massive bribing is taking place, constantly. The said salesman, however, has never once had to dirty his hands with a handful of bills in thirty years of working life (mini bribes, perhaps). And for one reason only: You don't have to bribe your *amicos*. He had made friends with so many purchasers that they gave him the commissions, if his offer was the best, even though he didn't give kickbacks—because you don't take money from friends. Sure: creating the kind of network that saves you a hundred thousand dollars in bribe money is not the easiest thing in the world.

The salesman would have had a much easier time had he simply given kickbacks. It's no coincidence that the word corrupt (from Latin corruptus) means, literally, "broken in pieces". Someone whose manager morale is so annihilated that they only want to coast lazily through their day, they'll have to resort to kickbacks. The aforementioned salesman is aware of the effort he has to put into his network, but it's worth it to him. "That little bit of effort is well spent", he says, "if I'm able to keep my conscience clear." If you want to keep a clear conscience, you'll find a way.

Corruption is not necessary when you put in a bit of effort. That's the core idea in compensation strategy: even when an order requiring a kickback would be huge—a good salesperson manages to more than compensate for this with "clean" orders. There is so much potential in every sale. Mining potential requires more work than an easy kickback. Corruption is always the easy way out. That's why we all have so much respect for colleagues who manage to stay "clean". Morality means effort. That earns our respect.

The marketing manager of a plant manufacturing company told me of another tactic. She turned the tables on the purchasers asking for kickbacks: "Either we get this order without a kickback, or I'll tell your board immediately how corrupt you are!" I don't necessarily want to recommend this strategy to those of you at home. If you bluff like that, you need to be pretty tough. But even if you don't have guts like that, you can still learn from this: that there is always (at least) one alternative to corruption. Corruption is something for slackers and the mentally dull and it will carry on *ad absurdum*. It doesn't define the person as a whole as bad. No: if you give in to corruption, you're just taking the easy way out. Good managers don't need to.

Moral, Practical and Good for You too

The crisis manager from our example above, by the way, didn't pay the six figure kickback. He lay awake five nights. During the day he set up the account and ensured that the transaction would a) not be discovered and b) would land in the pocket of the general and not that of the lieutenant. On the sixth night he finally made a decision—and slept like a baby.

Only a few hours earlier he had met with an ex-board member friend at a local wine bar. "Michael", said the friend getting right to the point, "don't touch that with a ten foot pole, even if you don't get your promotion to the board—which I don't think will be the case, our board isn't that corrupt. Look at me: I paid a bit now and again. Not much, but even that little bit robs me of sleep from time to time in my retirement years. That will follow you forever. I would pay a lot of money to get rid of that mistake. And I don't know anyone in my position who doesn't feel the same. None of us is that callous. We have consciences, too, you know."

The crisis manager took things one step further: He not only did not become corrupt, he told his board about the corruption, and took a stand: "I'm busting my chops here, but I'm not going that far. If that comes out, it will ruin our company's reputation. If you send me out to 'no man's land' because I haven't achieved my goals, then so be it. I want honest success and not this compromise to my character." His bosses couldn't do anything other than grit their teeth and show their respect. The manager had to wait two years, but he eventually made it to the board. Without compromising his character.

He also confided to me: “I have a son. How can I be a role model for him, if I let myself be corrupted?” *Ergo*: If you stop concentrating solely on business, and take a little peak outside the box, you won’t be such an easy target. Of course, there will always be weak, corruptible people, that’s not the point. The only important thing is: if you don’t want to compromise, don’t. Why should you?

Why Bother Being Honest?

I want to be honest. Every time I feel tempted, I’m thinking: “Man, that would really help me a lot right now. I could buy some great stuff with that money!” Top consultants don’t make as much as most people imagine. And you can never have enough money, right? But I was a banker long enough to know that every account has two sides. So I do myself a favor and create a balance sheet for myself.

Before you give in to temptation: Create a balance sheet.

Up at the top of the page I put the word “conscience”. Don’t laugh, I know “conscience” was replaced by “price mechanisms” as the new moral imperative ever since neo-capitalism was invented. So let me phrase that a bit differently: What price are you paying for corruption? The first factor is most likely not a financial one:

I know a psychologist who doubles as a coach to top executives, now and then. She confided to me: “Many carry on with their unreported earnings and tax evasion year after year. However, it always caught up with them at a moment when they least expected it, or weren’t able to deal with it. One “unfine” day, stigmatized, ostracized and alone, they learned that money, power and status don’t really make you happy. Then they experience an existential crisis that I wouldn’t wish on my worst enemy.”

Consider it this way: If a young manager takes and gives something—that’s one thing. We were all once “young and needed the money" (for our ego, seldom for a new refrigerator). But if you still think at the age of forty-five that money makes the world go round, then good luck with that (and your retirement –your conscience won't let you enjoy it). What I'm trying to say is:

If you don't compromise your values, you're doing it for yourself, your ego, your self-confidence, your code of honor and your soul!

I can usually go into a board meeting and know instinctively within five minutes who is holding their hand out and who isn't: managers who don't compromise their values have backbone and demonstrate authenticity, quiet authority and an integrity that deeply impresses the others. This puts the wind in their sails and gives them strength. You, yourself, have the most to gain when you remain honest.

The Curse of Every Evil Deed

Believe me: If corruption really paid off, I'd jump on the band wagon and endorse it, too. The problem is: it doesn't. Even if no one ever finds out, it still isn't worth giving in to temptation.

Tickets to tv shows, for example, are a typical bribe for lower level managers. A young, single-parent saleswoman received tickets for herself and her children for a popular tv show from a telecommunications company whose offer she was currently considering. Of course, "completely without ulterior motive". A saleswoman with two children is not exactly swimming in money and can afford to take her kids to such a show when it snows in July. She said to herself: "If I don't take the tickets, they'll be wasted, or someone else will get them. Everyone takes something! It's only a tiny amount, and I'll definitely stay objective in my decision for or against a certain service provider. I can separate my business from my personal life." That's pretty much the rationalization everyone uses when they get such an offer.

When the saleswoman asked me in a rather roundabout way whether she should accept the tickets, I asked her back: "Who is it going to cost the most?" She looked at me hard. She thought about it for a second, then said: "I'm giving the tickets back. No question." Then she thanked me. I still don't know what costs she was thinking of. Perhaps the misappropriated funds of the company's customers. It doesn't really matter, what's important is:

Every instance of corruption has its costs. Before you take or give, look more closely at the costs and who's paying!

Don't forget to factor in the benefits of abstention. I asked the young mother: "What do you gain by giving back the tickets? What great story will you be able to tell your two daughters this evening?" She smiled. "That mom stayed clean." How much money is that worth?

Furthermore, I recommend everyone who is led into temptation to consider the second line of Friedrich Schiller's famous quote:

What is the curse of every evil deed? That propagating still, it brings forth evil.

I know a director who savors the pleasures of her position's privileges. Beautiful car (Does she really need the S-Class?) handsome assistant, amazing "business trips". Her working time is used, quite naturally, indulging in her "on-the-side" hobby, real estate. Countless other top managers spend countless more hours managing projects for their clubs and associations in addition to those of their companies.

The ex-boss of an Austrian private bank was reported to have held, I kid you not, 153 different offices in addition to CEO, supervisory board, etc., etc. Presumably, his various offices didn't even relate to his own company. When the good man spent an average of one day per year for each of those functions, and still enjoyed a few days of vacation—when did he have time to lead his own company?

Another board member, a former engineer, spent half a day playing around with pet research projects which should have been the work of his research and development staff. All these little acts of self-enrichment are the message in the curse: they propagate further evil. The employees of the development-happy board member also only worked half a day: "When the boss can spend his days on his hobby, so can we!"

Corruption is infectious and shit doesn't sink. If the guys "up there" are helping themselves to a piece of the pie, so will the guys "down there". That's how you ruin organizations.

Because people are basically selfish (Adam Smith said it's what capitalism is based on), one could also say: If it's good for me, why not go ahead and ruin the company? But consider this: while it might be good for "me", what if there's something better?

What's Better than Corruption?

I have always been interested in whether corruption delivers what managers are hoping for. Every executive I've ever seen who was on the take, or giving kickbacks, I compared to the ones in that company or sector that remained clean. Guess which ones were happier?

The corrupt seem so driven. That makes sense because if you're always on the take, there must be a huge inner drive at work. For notorious bribe givers, career is usually everything. They're like junkies: if they don't get their fix, their hands begin to shake. Corruption is the drug (crime) of management junkies.

Corruption has a lot in common with the biblical sins. Lust, gluttony and drunken revelry may seem "totally awesome" at the time, but later you feel somehow dirty, demeaned, or empty. People who don't give in to temptation miss out on the quick thrill, but, in general, seem more healthy, happy, relaxed, balanced, respected, and at peace with themselves and the world.

A solid system of values will make you happier than the short-lived thrill of sin.

Which values? When I look at the ones who stay clean, the following spring into mind: foresight, endurance, determination, diligence, reliability, honesty, openness, resolution, compassion towards employees and a responsible attitude towards society.

It's obvious that it's a lot more difficult to live according to these values than to hand over, or pocket, a thick wad of cash. But it's massively more rewarding. What a shock it is for those of us in the post-modern world:

A moral life is a much more rewarding life.

Morality is not a luxury, it's the key to a fulfilling existence.

Vision not Corruption

If morality is not your thing, you should consider the following:

Corruption seldom makes dreams come true. Visions do.

Corruption is never an end in itself. Corrupt managers are chasing a goal that will lead them astray: its sweet lies promise as much success as a bridge built with chewing gum—it won't hold. The above-mentioned development-happy board member wanted to fulfill his dreams by "playing hooky" at work. He put in several hours each day, yet his guilty conscience—this damn place!—plagued him increasingly. So he gave up his position on the board, bought a young company, and spends 24 hours a day developing his heart out. He's as happy as a little boy.

That is the reason why some top managers "opt out" midway in their careers to do something completely different: start up their own company, go into consulting, volunteer, head up a local club, lead third world projects or become teachers. They have realized that not only corruption corrupts anything that doesn't serve their vision is compromising. Have you asked yourself today: What makes me really happy? You're truly blessed if you give yourself the luxury of this little question. Those who have suppressed this question their entire lives, who were "just" an average citizen, an average manager or an average husband ... they are roasting in the eternal fires of damnation.

If you follow your heart's vision, you don't need the crutch called "corruption". If you can fly, you don't need a cane.

The Chapter at a Glance: Take the Hours Test!

Ask yourself every hour: Is my behavior in line with my values? And: What makes me really happy? What will I do in the next hour to stay true to my vision?

Afterword—Taking Measure

"Hercules left his shepherds and fold to repair to a secluded area and reflect over what course his life should now take."

Gustav Schwab, Legends of Classical Antiquity

You're still there? Great! That makes me feel good. How did you like the book and all the personal stories in it from myself and the lives of other managers? This book is different from most business manuals and text books. And I like that (not all too surprisingly).

I was a manager long enough, and knew plenty of executives, to know that business textbooks are bought, but seldom read, and even less often implemented. The ancient Greeks knew that, too. That's why they didn't write self-help books or business treatises. They wrote myths and epics like the Odyssey and Iliad. They knew that people learned best from the stories of other people. They educated themselves through personal contact and passing on stories. They listened politely to experts and paid them, willingly, too—but they seldom learned from them. Sometimes I think a few of my customers are paying me for my consulting because they don't want to have to learn.

When clients of mine acquire new knowledge and are able to put it into practice, it's not because my change strategies were so good (they are good), but because I told them the right story, ideally from their own company. Stories like those you have just read. Such field reports change companies. That's why there is so much story telling in management. And I bet you learned more in the pages you just read, than if I had dressed it all up in text book form—and you (and I) wouldn't have had so much fun. C'mon admit it, you smiled a lot less when reading Peter Drucker.

That is not the *idée fixe* of a crazy ex-board member. In my opinion it's a sign of the times: we put the experts up on a pedestal. We listen to these so-called "gurus", but seldom take anything away that we can really use (at least for very long). We have unlearned what we long ago developed into "Action Learning" and "Appreciative Inquiry": learning from ourselves, from friends on the board, the good or corrupt colleagues, customers, employees, spouses, children, grandma and grandpa, the guy at the newsstand. Just as the Indians did around the bonfire: sitting together and exchanging stories from today, so that tomorrow is better.

Almost everyone I asked to read a sample chapter of this book, said "I can completely relate to that!" I was hoping for just that kind of reaction. Do you know what people usually say after reading your average business book? They say: "Hmm, interesting, I should really try that." And then they don't, because they can't fit it in with their own situation. Publishers in Germany weren't exactly thrilled with my concept of experience-led learning. Most wanted to shelve it, immediately, because it wasn't paying homage to the cult of the expert. Luckily, there are publishers who put intellect before cult.

Let's wave goodbye to a culture of experts. It's because of them that we've gotten into such a fix. Or are you happy with the today's "throw away" mentality? We need a more human (learning) culture, where real people (and their experiences) and not experts with their abstractions, are the focus.

That's why I want to encourage you to continue along the path that you've begun here: take a look at what's happening around you. You'll find teachers everywhere! You can (and should) learn from everyone and everything that you come into contact with. This kind of learning is far more effective than that conducted by experts, which is comparable to the LA Lakers playing against a Venice Beach pick up team (nothing against street ball at Venice Beach). Learn from your own example and experience, and that of others. And set an example yourself. Relate your own experiences: you're not only doing your own personal and business development a great service, you're also helping your listeners. And me too, by the way, because where do you think I got all these great stories from? Share your best and worse, funniest and most stressful or annoying "downfalls" with me. Here, at www.klausschuster.eu.

Acknowledgements

I want to thank all the big and little sinners in management who have confessed their trivial and deadly sins to me with astonishing openness and honesty—well knowing that I would be breaking the seal of confession with the publication of this book. I also want to thank my parents for their attentive vigilance, above and beyond writing this book—without their help in keeping my feet on the ground, I could have flown, like so many top managers and Icarus, into the annals of ignominy and shame. My wife, Jana, is due my eternal thanks for her patience and, above all, the willing support she has given me and my craziest ideas (such as writing a book). And to our two daughters, Teja and Tina, the pillars of virtue in our household, go my thanks for their relentlessness and regularity in pointing out my own personal major and minor sins in family management. My golden retriever, Lan, won't benefit from my literary thanks (since he doesn't like to read that far in a book)—nevertheless, I am forever in his debt for the regularity of our walks which have saved me from an author's inevitable back pain and expensive chiropractor bills.

About the Author

Klaus Schuster, MBA, was an executive board member for many years at a large international financial institution. There he acted as troubleshooter, constantly travelling, and spearheading the set up of a new company branch in Eastern Europe. In the meantime, he has founded his own company which advises, coaches and trains top managers and junior executives in all industries and sectors. He is a highly regarded international author, who writes business articles and columns in Europe. Recently, he has been making the headlines in Europe with the enormous (and thankless) task of cleaning up the messes of a middle European bank on behalf of the national bank and European Union. "11 Management Sins that You Should Avoid" is the first of his four, very successful, management books to be published in English.

Index